P9-DWP-618

SPAM

INFRASTRUCTURES SERIES

EDITED BY GEOFFREY BOWKER AND PAUL N. EDWARDS

Paul N. Edwards, *A Vast Machine: Computer Models, Climate Data, and the Politics of Global Warming*

Lawrence M. Busch, *Standards: Recipes for Reality*

Lisa Gitelman, ed., *"Raw Data" Is an Oxymoron*

Finn Brunton, *Spam: A Shadow History of the Internet*

Nil Disco and Eda Kranakis, eds., *Cosmopolitan Commons: Sharing Resources and Risks across Borders*

SPAM

A SHADOW HISTORY OF THE INTERNET

FINN BRUNTON

THE MIT PRESS
CAMBRIDGE, MASSACHUSETTS
LONDON, ENGLAND

MIT Press books may be purchased at special quantity discounts for business or sales promotional use. For information, please email special_sales@mitpress.mit.edu or write to Special Sales Department, The MIT Press, 55 Hayward Street, Cambridge, MA 02142.

This book was set in Engravers Gothic and Bembo by Toppan Best-set Premedia Limited, Hong Kong. Printed and bound in the United States of America.

Library of Congress Cataloging-in-Publication Data

Brunton, Finn, 1980–
Spam : a shadow history of the Internet / Finn Brunton.
pages cm.—(Infrastructures)
Includes bibliographical references and index.
ISBN 978-0-262-01887-6 (hardcover : alk. paper)
1. Spam (Electronic mail)—History. I. Title.
HE7553.B78 2013
384.3'4—dc23
2012034252

10 9 8 7 6 5 4 3 2 1

This book is dedicated to the memory of Gunard Solberg, dear friend and inspiration, deeply missed.

CONTENTS

ACKNOWLEDGMENTS

This book owes its start as a dissertation to Christopher Fynsk and Mario Biagioli. Without their advice, encouragement, feedback, provocations, and patience, it would never have been written. James Leach and Timothy Lenoir were kind enough to sit on my dissertation committee, where they provided excellent comments towards that document eventually becoming this book. Its production would likewise have been impossible without the conversation and support of Helen Nissenbaum while I was at NYU; portions of this book were written at NYU while funded by grants from the National Science Foundation (CNS/NetS 1058333) and the AFSOR-MURI Presidio grant. Paul Edwards, Gabrielle Hecht, and the STS Reading Group at the University of Michigan in Ann Arbor read a full draft of this manuscript and gave a writer's dream of thorough, exacting, and unfailingly helpful comments and advice. Finally, this book would not have taken shape without the work of Marguerite Avery, Katie Persons, Kathleen Caruso, Nancy Kotary, Celeste Newbrough, and the reviewers at MIT.

Many of the ideas, examples and conclusions in this book evolved while in conversation with Lucy Alford, C. W. Anderson, Gus Andrews, Burhanuddin Baki, Don Blumenfeld, danah boyd, Mia Sara Bruch, Douglass Carmichael, Michael Cohen, Sande Cohen, Gabriella Coleman, Jean Day and the reviewers at *Representations*, Charlie DeTar, Alexander Galloway, Kristoffer Gansing, Lisa Gitelman, Matt Gleeson, Sumana Harihareswara, Sean Higgins, Micah Jacob, Ben Kafka, Daniel V. Klein, Jorge Ledo, Joanne McNeil, Ben Marsden, Jay Murphy, Kathryn Myronuk, Quinn Norton, Camille Paloque-Bergès, Jussi Parikka, Leonard Richardson, Nora Roberts, Erica Robles, Tony Sampson, Luke Simcoe, Fred Turner, and a number of people who would like to remain anonymous. Thanks so much to you all. Any mistakes are, of course, my own.

INTRODUCTION: THE SHADOW HISTORY OF THE INTERNET

PRELUDE: THE GLOBAL SPAM MACHINE

Pitcairn Island in the South Pacific is the smallest and least populated jurisdiction in the world, described by the first mariner to observe it as a "great rock rising out of the sea . . . about a thousand leagues westward to the continent of America."[1] It's a lonely place; when the *Bounty* mutineers needed a refuge from the global empire whose laws they'd broken, they sailed for Pitcairn and vanished for almost two decades. As of last year's electoral rolls, there are 45 people on Pitcairn, most of them descended from the mutineers. There is a government-subsidized satellite Internet connection for the island's houses. Per capita, Pitcairn Island is the world's number-one source of spam.

How is this possible? Can this island, whose major export industries are handicrafts for passing ships and stamps for philatelists, host a cabal of spammers bombarding the world with phishing messages and Viagra ads by satellite connection? In the list of spam production per capita, Pitcairn is followed by Niue and Tokelau, also in the South Pacific; the Principality of Monaco, whose population overwhelmingly consists of rich expatriates dodging taxes in the world's second smallest country; and the Principality of Andorra, a country one-fifth the size of Rhode Island.[2] Are there really that many Catalan-speaking Andorran spam barons, accumulating stolen credit card data by the ski slopes of the Pyrenees?

The answer is no, of course. The Polynesians who live on Niue, like the wealthy Europeans overlooking the Mediterranean, have been unknowingly conscripted into the business of spam. Virtually all of us on the Internet have been, on one side or another. Spam has changed laws and communities at the points of friction where the network's capacities rub against prior orders of work and governance. It has changed our language, economics, and culture and exerted a profound effect on our technologies. It has subtly—and not so subtly—deformed the shape of life online, pulling it into new arrangements that make no more sense than the movement of the galaxies unless you allow for the mass of all the dark matter. Only at scale, in time and in space, can we begin to bring this shape into focus. It stretches from embryonic computer networks in the 1960s and 1970s to the global social graph of 2010, from the basements of MIT to the cybercafés of Lagos, Rotterdam, and Tallinn. It encompasses points across the network from covert chat channels to Google's server farms to "ghost number blocks" to anonymous banks of airport pay phones. The spam most of us experience every day as a minor and inexplicable irritant is like the lonely rock that sailor sighted, merely the visible peak of vast and submerged infrastructures with much to say about the networked world.

The word "spam" means very different things to different people at different times. It is a noun, collective and singular, as "this spam" can mean "all these messages I've received" or "this particular email." It is a verb, as in "they spam me," and an adjective, as in "this is spammy." It refers to many varieties of exploitation, malfeasance, and bad behavior, and spam terminology has branched out into specific subdomains, from "phishing spam" and "419 spam" to splogs, linkfarms, floodbots, content farms. (All of these new terms and forms will be defined and described in the following chapters.) It can easily slide into what philosophers call the "sorites paradox" ("sorites," appropriately for talking about spam, comes from the Greek word for "heap" or "pile") to describe the linguistic confusion between grains of sand and "sand" in dunes, the moment at which individual bits become a single big pile. When "spam" is discussed in journalism or casual conversation, the word is often meant as I have used it in the previous paragraph, as a unified substance or a continuous event, like smog or "a mass or pulp," as Susanna Paasonen puts it.[3] But spam begins to make sense only when we get specific and separate out the different types, motives, actors, and groups.

Spam is not a force of nature but the product of particular populations distributed through all the world's countries: programmers, con artists, cops, lawyers, bots and their botmasters, scientists, pill merchants, social media entrepreneurs, marketers, hackers, identity thieves, sysadmins, victims, pornographers, do-it-yourself vigilantes, government officials, and stock touts. Long as it is, this list mentions only those who live there more or less full-time, because everyone online participates in the system of spam, however indirectly. We fund and enable it with choices we make and trade-offs we are willing to accept because the benefits to the network outweigh the costs. We generate areas of relevance and attention whose discovery and exploitation is the heart of the business. We alter how spam works and break up its current order with choices, refusals, and purchases whose consequences we may not understand.

Those houses on Pitcairn, for example, connected to their satellite link, do not shelter spammers hard at work. At some point, a few of the computers on Pitcairn were taken over by a piece of malware, probably arriving as the misleading payload of a spam message that appeared to come from a friend, taking advantage of unpatched or insecure software that can be daunting for the user to maintain. This malware quietly commandeered the computers without their owners ever noticing (perhaps they merely thought that the machine, and the Internet connection, had become a little slower) and enrolled them along with many thousands of other captured computers in homes and offices around the world into a system called a "botnet." One of the functions of the botnet is to use spare computer power and Internet connection bandwidth to send out spam messages on behalf of the botnet's controller, who can thus generate hundreds of millions of messages at effectively no cost. A desktop computer, manufactured in Shenzhen or Xiamen or Chennai, powered by a diesel generator and online through a satellite over the Tropic of Capricorn, on a remote island in the Melvillean Pacific, has become part of a distributed machine controlled by a remote group of botmasters in Denver or St. Petersburg. This is a system unprecedented in human history, a vision out of science fiction, that writes constantly repeating messages of crushing banality: "YOU HAVE WON!!!/Congratulations!!! You have won £250,000.00POUNDS from Pepsi Company award 2010, Please provide your Full name, Age, Sex, Country, Occupation"; "Permanent En1arged-Penis/She prefers your lovestick bigger"; "Listen up. I must let you in on a few insider ⋆secrets⋆: Instead of waiting months to generate sales on your site, you can start gaining the hits you want right now."

THE TECHNOLOGICAL DRAMA OF SPAM, COMMUNITY, AND ATTENTION

This is a book about spam for anyone who wants to understand what spam is, how it works, and what it means, from the earliest computer networks to the present day. To understand spam means understanding what spam is *not*, because—as you will see—the history of spam is always a history of shifting definitions of what it is that spam harms and the wrong that it produces. The history of spam is the negative shape of the history of people gathering on computer networks, as people are the target of spam's stratagems. It is defined in opposition to the equally shifting and vague value of "community." (In fact, many of the early cases of spam provoke groups of people on computers into the task of self-definition and self-organization as communities.) To put this history and this book into a single sentence: spamming is the project of leveraging information technology to exploit existing gatherings of attention.

Attention, the scarce resource of human notice, is what makes a community on the network, and the creation of communities, the invention of "we" on the Internet, is an act of attention. Communities and spam as a whole are projects in the allocation of attention, and spam is the difference—the shear—between what we as humans are capable of evaluating and giving our attention, and the volume of material our machines are capable of generating and distributing when taken to their functional extremes.[4] Over four decades of networked computing, spammers have worked in that gulf between our human capacities and our machinic capabilities, often by directly exploiting the same technologies and beneficial effects that enable the communities on which they predate. These two forces build, provoke, and attack each other, and the history of one cannot be understood without accounting for its nemesis.

This co-constitutive feedback loop between communities and spammers is a major event in the technological drama that is the Internet and its sister networks. This term, "technological drama," is from the work of Bryan Pfaffenberger and provides framing and context for the story to come; it is the life of a technology, from conception and design through adoption, adaptation, and obsolescence.[5] Why a *drama*? Because technologies are statements about the distribution of needful things such as power, status, access, wealth, and privilege, for good and for ill. They are stages on which social and political arguments and counterarguments are made. These arguments may be not merely irrational but also deeply cultural and ritualistic, expressing convictions that lie well beyond the immediate physical constraints of engineering. A technological drama is suspenseful: it is not played out to foregone conclusions but is rife with the possibility of subversion, takeover, and unexpected captures. It is a drama of escalation and feedback loops, for technologies are never merely passive vessels for holding ideas and ideologies but active things in the world that open new possibilities and capacities. They change the communities that created them and those that take them up.

The inciting incident that frames any technological drama and gets it moving is the gathering of a "design constituency," in Pfaffenberger's phrase. "Constituency" is well-chosen, because we are not simply referring to the engineers, inventors, scientists, or designers who actually work the lathe, draw out the blueprints, or blow the glass but to the entire body of people who participate in the design and who stand to directly benefit from the technology's success. It is to their benefit if it sells, of course, assuming that it is the kind of technology that can be commoditized into widgets and not, for instance, a civil engineering project. More important, however, is that the values embedded in the technology, intentionally or unintentionally, become dominant. Those values reflect an arrangement of power, control, and prestige that the design constituency would like to see in the world, whether centralized and privatized, open and egalitarian, or otherwise. This design constituency can include the engineers and applied designers themselves, as well as managers and shareholders in firms, but also politicians, experts, theorists, and elites. What is complex and important here is to be able to view technologies in two ways at once: seeing both their physical capacities and functions, and their social and political assertions—the moves they make in the allocation of resources and power. We will explore computer networks and the people building and using them with this perspective.

To get some perspective on how technological dramas work as an analytic tool, consider the case of heavier-than-air aviation. No one would argue against the profound benefits delivered by the development of powered flight, but to really understand the adoption and adaptation of the technology we must account for the acts of political stake-planting and the renegotiations of influence and control that went with it. Aviation's roots included the powerful early idea of "air-mindedness," for which planes were not just powered, winged craft for flight but objects whose existence was a direct expression of a rational, modern, and global mindset that would simultaneously bring about world peace and subdue colonized populations through bombing.[6] H. G. Wells, for instance, in his role as a public intellectual and policy advocate, framed the work of getting from one place to another by air as nothing less than the formation of a new generation of intellectual pilot-samurai who would enforce a technocratic world state. This coming class would be necessarily scientific, cosmopolitan, and forward thinking, because they used planes. Airplanes, Le Corbusier asserted, were "an indictment, an accusation, a summons" to architects and city planners, shaming their retrograde ideas: planes were pure futurity, the avatars of the machine age.[7] At the same moment, figures such as Gabriele D'Annunzio, aviator and Fascist literary icon, and Italo Balbo, commander of Italian North Africa and leader of the famous Century of Progress Rome–Chicago flight, were using both the glamour and threat of flight as an assertion of Fascism's fusion of futuristic dynamism with nationalist and archaic agendas. In the United States, flight included the democratic vision of self-taught tinkerers hacking on planes in barns and garages and potent assertions of military superiority and the projection of power abroad in an abruptly smaller world. And so on. This many-layered complexity of motives, ideas, fantasies, and goals obtains across the technological domain. To understand any technical event in depth, we need to be able to describe the full range of interests in the work of the design constituency.

The other side of the drama is the "impact constituency," those whose situation is being changed by what the design constituency produces. Neil Postman simply called them the "losers," because the rearrangement of the world, although not necessarily one in which they lose out explicitly, is one over which they have little initial control. Examples from Pfaffenberger's research are particularly well suited for this purpose. Consider the politics of irrigation projects in Sri Lanka, a particularly bald-faced attempt at social engineering via technological deployment. Landed, powerful, and deeply anti-industrial local elites sought to manage the threat of dispossessed, dissatisfied, and mobile peasants by creating a neatly controllable system of rural settlements. It was a project of getting them back on the farm, to put it simply, where they would stay ignorant of modernity and easier to tax and manage, forestalling the arrival of an industrial order in which the landholding "brown sahibs" stood to lose. The landholders did the designing, and peasants felt the impact. James C. Scott's work provides several cases along these lines—of compulsory villagization, collectivization of agriculture, and high modernist urban planning—in which the capture and redistribution of value is exerted through the production of artifacts, places, and systems of living, always framed as "for their own good."[8] We could also speak on the far smaller and more intimate scale of something like the Victorian household's hallway bench, with its ornate mirrors and carved hooks, for the household's masters to view as they walk by and hang their hats, and the strikingly uncomfortable bare low bench on which the servants, messengers, and peddlers are to sit and wait.

What makes these relationships of design and impact into a *drama* is the back-and-forth of technological statements and counterstatements. After the original assertion made in the design and deployment of a technology come the responses, as the impact constituencies take things up, change them, and accept them or fight back. The design constituency cooks up methods and language for using technologies to arrange, distribute, or contain power, status, and wealth, and impact constituencies have countermoves available. They can produce arguments to get their hands on the technology, for instance, and reconstitute it, which does not merely ameliorate the painful setbacks produced by the deployment but actually generates a new technology that builds on the old for their own purposes.

The most obvious and canonical instance of reconstitution in a technological drama, producing a "counterartifact" out of the existing technology, is the personal computer. Decades of development in computing had been the product of the military, academia, and contractors and corporations such as IBM. The computer was "the government machine," sitting in the ballistics lab and running models for building nuclear weapons and game theory analyses of Cold War outcomes.[9] Computing had become virtually synonymous with a bureaucratic, administered society in which people were subsumed as mere components, in what Lewis Mumford termed the "megamachine." Against this concept came the first hackers and subversive tinkerers, activists, and artists, with Ted Nelson (who coined the term "hypertext") asserting, "You can and must understand computers *now*," "countercomputers" for the counterculture, the Homebrew Computer

Club, the Apple I and II, and so on.[10] The drama never stops, and Fortune's wheel keeps turning, casting down and raising up and demanding new technical statements and counterstatements.

Getting a significant new technology that is instilled with cultural values and political goals to "work" is an act of assembly, of mobilizing different groups, ideas, forces, and stakeholders together. There needs to be a flexibility to the idea, enough that you can pull heterogeneous and sometimes even opposed powers into an alliance behind it. A technology too clearly and narrowly defined may not be able to develop all the alliances it needs. For the Sri Lankan irrigation project, it meant creating a complicated alliance of ethnic and religious chauvinism, paternalistic philanthropy, opposition to the old British colonial order (by an elite whose power derived from the restructuring of society under the British), and so on. A similar multiplex set of alliances could be seen in the free/libre open source software (FOSS) movement, with businesses, individual charismatic activists, developers, and political radicals of many stripes and very diverse agendas trying to gather under the same banner. The epic semantic fork between the models of open source and free software in 1998 captures one of the moments when the ambiguities became unsustainable and had to be reformulated. "Movement," writes Christopher Kelty, "is an awkward word; not all participants would define their participation this way. . . . [They] are neither corporations nor organizations nor consortia (for there are no organizations to consort); they are neither national, sub-national, nor international; they are not 'collectives' because no membership is required or assumed. . . . They are not an 'informal' organization, because there is no formal equivalent to mimic or annul. Nor are they quite a crowd, for a crowd can attract participants who have no idea what the goal of the crowd is."[11] This complex mesh, sharing practices and debating ideology, is a design constituency, gathering around the technology and trying to marshal support for different and sometimes conflicting visions to push the project into the world.

This process of gathering and marshaling is strengthened by some founding ambiguities. These let the designers cast a wide net and make it easier to argue that this technological project speaks for *you*, too. Some of these ambiguities are invoked of necessity. The story that a design constituency builds to draw in stakeholders from different domains to support a new technology must draw on what Victor Turner terms "root paradigms," the deep organizing principles of a culture and an epoch that provide a rationale and a motive for action. Root paradigms aren't exact and precise, and they are never simply true or false. Whether they be submission to the free market, the sanctity of human life, the vital and cleansing power of war, or the unquestionable role of the dynastic king, root paradigms are dynamic, messy, and enormously powerful concepts built on internal oppositions. They draw their energy and vitality from their unsettled condition of irreconcilable struggle within which new technologies, political initiatives, and movements can be placed and contextualized. At major turning points in the development of the Internet and spam, struggles between constituencies were played out that drew on far older root paradigms such as absolute freedom of speech, communal self-defense and self-organization, the technological autonomy of the capable individual, the inevitability of destructive anarchy without governance, and the centrality of commerce to society. The presence of these paradigms gives technological deployments the thrilling, and often later inexplicable, attraction of social movements (because that is, in fact, what they are). They draw their strength from roots sunk deep into the earth, where the bones of prior orders and the ruins of previous civilizations underlie the present.

These foundational ambiguities in a technology's design are a crucial resource for the impact constituencies and others to exploit. Sally Moore describes how a reworking of the arrangements introduced by a technology is made possible, by "exploiting the indeterminacies of the situation, or by generating such indeterminacies . . . areas of inconsistency, contradiction, conflict, ambiguity, or open areas that are normatively indeterminate."[12] The indeterminate space is the place for trade-offs and concessions, ways to get many diverse parties working together and pointed in the general direction envisioned by the design constituency. It is also leaves apertures and affordances in the plan for the manipulation, escapes, and exploitation of others, from innovations and improvements to exploits and deliberate sabotage—and thus for things like spam.

This complex indeterminacy obtained at every stage of the Internet's development. As will be discussed later in this book, there was deep uncertainty and widely varying understandings as to what this thing was to be and how it should be used by people and by hardware and software. This uncertainty was an enormous boon for innovators and inventors, for the strange frontiers of network culture, and for both hackers and criminals, whose somewhat blurry relationship and ambiguous legal status recurs in this history. Spam survived and prospered by operating in the edge cases around these big ideas, in the friction between technical facts and the root paradigms that are expressed in them where powerful concepts like trust, anonymity, liberty, and community online were reinvented, modified, and sometimes discarded. In following spam, we will explore how these ideas evolved and, above all, how human attention online, and the strategies for capturing it, changed over time.

THE THREE EPOCHS OF SPAM

Appropriately for a technological drama, the history of spam has three distinct acts, which are reflected in this book's three sections. The first, from the early 1970s to 1995, begins with conversations among the architects of the earliest computer networks, who were trying to work out acceptable rules, mores, and enforcement tools for online communication. It closes with the privatization of the Internet and the end of the ban on commercial activity, the arrival of the Web, and the explosion of spam that followed the Green Card Lottery message on Usenet in May 1994. It is a quarter-century of polylogue concerning the fate and the purpose of this astonishing thing the online population was fashioning and sharing ("polylogue" being a term from an early computer network for this new form of asynchronous and many-voiced conversation on screens). It includes a remarkable cast of postnational anarchists, baronial system administrators, visionary protocol designers, community-building "process queens," technolibertarian engineers, and a distributed mob of angry antispam activists. It is an era of friction between concepts of communal utility, free speech, and self-governance, all of which were shaped in a negative way by spam. "Spam" here is still used in its precommercial meaning of undesirable text, whether repetitive, excessive, or interfering. The imminent metamorphosis of these ideas as the values and vision of the network changed in the mid-1990s was partially signaled and partially led by changes in spam's significance and means of production.

The next phase lasts about eight years, from 1995 to 2003, or from the privatization of the network through the collapse of the dot-com boom and the passage of the CAN-SPAM Act in the United States. It is about money and the balance between law and collective social action. Those years are filled with the diversification of spam into a huge range of methods and markets, following and sometimes driving innovations on the Internet and the Web, from search engine manipulation and stock market "pump-and-dump" schemes to fake password requests and "Nigerian prince" fraud campaigns. During this time, a strange class of magnates and hustlers is born, arbitraging the speed of new technological developments against the gradual pace of projects to regulate them. Their nemeses start here as well, with antispam projects ranging from message-filtering programs to laws and coordinated campaigns of surveillance, research, and harassment. This period is fraught with uncertainty about the role of nations and territorial boundaries online, the ambiguous rights and responsibilities of "users," and the relationship between what algorithms can process and what humans can read.

The most recent phase, from 2003 to the present day, turns on these questions of algorithms and human attention. A constellation of events is dramatically changing the economics of the spam business: the enforcement of laws, the widespread adoption of powerful spam filters, and the creation of user-produced content tools. To keep the business profitable, those spammers who survive the transition will develop systems of automation and distributed computing that span the globe and lead to military applications—building nothing less than a criminal infrastructure. In turn, antispammers will rely on sophisticated algorithms and big data to minimize the labor of catching and blocking spam messages. As spam prefigured and provoked crises in community and governance on the Internet, it now shows us an imminent crisis of attention—in the most abject and extreme form, as always. After four decades of interrupting conversations, grabbing clicks, demanding replies, and muddling search results, spam has much to teach us about the technologies that capture our attention, now and to come.

I READY FOR NEXT MESSAGE: 1971–1994

SPAM AND THE INVENTION OF ONLINE COMMUNITY

Life will be happier for the on-line individual because the people with whom one interacts most strongly will be selected more by commonality of interests and goals than by accidents of proximity.

—J. C. R. Licklider and Robert Taylor, 1968

Those who buy into the myth that Cyberspace is a real place also believe that this illusory locale houses a community, with a set of laws, rules, and ethics all its own. Unfortunately, the perceived behavior codes of Cyberspace are often in conflict with the laws of more substantive lands like, for instance, the United States of America.

—Lawrence Canter and Martha Siegel (creators of the green card lottery spam), 1995

GALAPAGOS

To understand the history and the meaning of spamming in depth, we need to understand the history of computer networks, the people on them, and the cultures they created. In many ways, the most remarkable thing about the word "spam" is its transitivity and portability. It is used to talk about completely different kinds of activity, not just on different websites or applications but on completely different *networks.* Forms of "spamming" existed on networks like Usenet and The WELL, quite distinct from the Internet, as well as on email and the web, and web-based systems from blog comments and wikis to Twitter. What do we mean, then, by a network of computers—the prerequisite for spam? It is something contingent and historically complex, something that could have developed in very different ways. There was not one network, but many. Spam, in all its meanings, is entwined with this polymorphic history.[1]

Initially, computer networks were specific to types of machines, institutions, and projects. The Internet, when it came, was the network of these networks, and early forms of spam retained some local character from the different kinds of systems on which they developed. In the beginning, the programmers would create an operating system specifically for a single computer, purpose-built for a customer's goals. Likewise, the engineers and architects built each network for its specific purpose, with an enormous diversity of approaches, and it is important to remember how many and how strange were the systems made so computers could talk together.

For a start, there were networks for businesses, such as the international packet-switching network (using the same technology underlying the Internet) for managing airline reservations in the late 1960s and the General Electric Information Services, part of GE's construction of their own continent-wide network for internal use. There were military defense networks, such as the Semi-Automatic Ground Environment or SAGE, fully operational in 1963, which could display all of U.S. airspace in real time so that the Air Defense Command could spot potential enemy planes and coordinate a response, using lightguns to "click" on the screen. It was a milestone of technological planetary visualization running for crewcut specialists in Mutually Assured Destruction, with an ashtray built into every console. Yet by the early 1970s, there were also hippie networks, such as Community Memory in Berkeley, which was a couple of teletype terminals—boxes with keyboards where a scroll of paper is the "screen"—in a record store and a public library for posting comments and questions and reading replies relayed from a refurbished banking computer hustled for free from a warehouse. Counter-computing for the counterculture:

```
> FIND PHOTOGRAPHY
1 ITEMS FOUND
> PRINT
#1: MELLOW DUDE SEEKS FOLKS INTO NON-EXPLOITABLE PHOTOGRAPHY/
MODELING/BOTH . . . OM SHANTI
```

And so on: where to find good bagels and a good bassist, plus rants, freakouts, somber conspiracy theorizing, Burroughs parodies, and much quoting of rock lyrics, a textual genre mash difficult to describe precisely but instantly recognizable to anyone who has spent time in online chat spaces and message boards.[2]

There were hobby networks once the price of PCs and modems had dropped enough to be affordable for a somewhat wider audience. Bulletin Board Systems or BBSes, set up with a phone line and disk drive, took local calls one at a time to upload and download files from a disk. These little isolated systems were akin to the farmer's telephone networks before the consolidation of the Bell monopoly, when phones interconnected a small town, transmitting over spare barbed wire and insulators made from the sawn-off necks of Coca-Cola bottles.[3] FidoNet was developed to let BBSes call each other and pass files and messages edgewise, from one board to the next, hooking up area codes, regions, the nation, and the world. Then came software that built Fido out further so that BBS users could post messages to common discussion groups shared among many BBSes, and the boards multiplied into the thousands and tens of thousands through the late 1980s and early 1990s.[4]

There was a network devoted to passing messages back and forth between IBM computers on university campuses, a research network for physicists, and a proprietary network protocol specific to machines made by the Digital Equipment Corporation (DEC), called DECNET (on which yet another group of physicists built yet another scientific collaboration network). There was an extremely meticulous and demanding network protocol called X.25 developed by the big telephonic standards bodies in the image of Bell's network—carefully run, end to end, with painstaking error correction—that underlay ATMs. X.25 was also used by some national networks such as Transpac in France, where it was the basis for the "French Internet" that was Minitel—millions of phone terminals with keyboards that could be used to look up phone numbers and buy airplane tickets and trade erotic chat messages (*messageries roses*) and coordinate a national student strike (as happened in 1986).[5] There were enormous struggles for adoption and market share between competing standards.[6]

As mentioned earlier, there was also The WELL (for "Whole Earth 'Lectronic Link"), which ran on an old DEC computer in Sausalito, California—an early social network hosting an ongoing billion-word conversation among the Deadheads, hippies, and early personal computer adopters of the Bay Area. (The WELL, SAGE, DECNET, BBS: geeks and the military are both very fond of acronyms, for which I ask the reader's indulgence for now.) There were games on some of these early networked computers that you could play with other people, games with names like "MUD1" and "TinyHELL" and "Scepter of Goth" and "FurryMUCK": new environments and settings for communication. There was a network that sent packets of data by radio around Hawaii because the phone lines were unreliable.[7]

With a few experimental exceptions, these networks could not talk to each other, and the users of one could not send messages or trade files with the users of another. It was an ocean scattered with Galapagian islands, only rarely communicating, with native forms evolving at their own speed and in their own direction, "most beautiful and most wonderful."

It was an ideal situation for a scholar of digital culture looking for convergent shapes and genealogies, but a nightmare for an engineer seeking efficiency and interoperability. This difficulty was exemplified by Room 3D-200 in the Pentagon in 1966: the office of the Advanced Research Projects Agency (ARPA). It was Robert Taylor's office; he was an experimental psychologist turned engineer who had recently left NASA for ARPA, and the suite next to his office was occupied by three bulky terminals.[8] Each gave him access to a separate remote computer mainframe: the Compatible Time-Sharing System (CTSS) at MIT, Project GENIE at University of Calfornia–Berkeley, and the computer at the System Development Corporation in California (which had taken part in the construction of the huge SAGE system described earlier). Users of the three networks could not talk across them or share resources between the mainframes. This setup was expensive and obviously redundant; it prevented the groups on the separate university systems from working together, and Taylor had strong feelings about collaboration. "As soon as the timesharing system became usable [at a university], these people began to know one another, share a lot of information, and ask of one another, 'How do I use this? Where do I find that?' It was really phenomenal to see this computer become a medium that stimulated the formation of a human community."[9] That was the effect of just one machine at one university switching to time-sharing from a batch processing system. A *batch processing* computer did tasks one at a time in order and, as computer scientist John Naughton put it, you only knew the other people using the computer as names in the queue taking up processing time.[10] Imagine, then, what hooking all the different time-sharing networks together might accomplish—the "'community' use of computers," in Taylor's words, assembling "a 'critical mass' of talent by allowing geographically separated people to work effectively in interaction with a system."[11]

A network of networks, for reasons that combined utility, efficiency, and budget with a concept of "human community": this was the vision for a new project, christened ARPANET, whose genesis and demise bracket this epoch of spam. The first recorded instance of spam behavior occurred on that CTSS network at MIT, and Taylor would have received it at his terminal in the Pentagon. The second protospam was sent on ARPANET itself. Because spam will appear, in all its forms, in relation to the community, each calling the other into being, we must understand what we mean by "community" on these networks.

THE SUPERCOMMUNITY AND THE REACTIVE PUBLIC

Taylor worked with J. C. R. Licklider, a Johnny Appleseed of computing who had been part of SAGE and ARPA and many other influential projects. Licklider was deeply convinced of the transformative power of the computer, writing visionary documents about "man–computer symbiosis" and addressing memos within ARPA to "Members and Affiliates of the Intergalactic Computer Network."[12] Licklider and Taylor coauthored the 1968 landmark paper "The Computer as a Communication Device," in which they envisioned, blooming on the network of networks, nothing less than the community of communities, making "available to all the members of all the communities the programs and data resources of the entire super community."[13] The supercommunity was like the supercomputer: the power of many minds, like many processors, running in parallel and efficiently distributing the load of thinking and working.

These two elements of computing infrastructure and social order were closely, rhetorically conjoined. "What we wanted to preserve," wrote the computer scientist Dennis Ritchie about losing access to an early time-sharing system at Bell Labs, "was not just a good environment in which to do programming, but a system around which a fellowship could form"—nothing less than "communal computing."[14] The system at MIT that Taylor was connected to from the Pentagon, CTSS, was a direct ancestor of an important system called Multics. Multics had a command called "who" that would tell you who else was logged in while you were. Knowing this information meant that you could contact someone who was also logged in with an electronic message or a phone call. It was not simply a computing system, but the starting point of online society. "The who command," writes the software engineer Tom Van Vleck, whom we will encounter again at the first appearance of spam, "contains the tacit assumption that the users of the Multics installation are all reasonable colleagues, with some shared set of values."[15] Or, as administrators Cliff Figallo and John Coate said of watching The WELL's devoted users face a social crisis: "We looked at each other and said, 'They're calling it a community. Wow.'"[16]

"Community" and "spam" are both difficult things to talk clearly about: outstanding examples of words as places rather than fixed objects, that is, zones where we can meet and negotiate about meaning. These words act as open space for the movements of great powers and agendas, as well as small roving groups of actors. "Community" enables conversations about its meaning. For early sociologists such as Ferdinand Tönnies and Émile Durkheim, community (along with the similarly spacious "society") makes room for describing the condition of people together after the advent of industrial modernity—indeed, for drawing opposing conclusions about that condition. For Jean Lave and Etienne Wenger, "communities of practice" are an area for theorizing learning and, later, knowledge management. For the Chicago School of sociology—and Marshall McLuhan, and those in his aphoristic wake—community, among other things, frames the conversation around theories of media, such as the visionary "Great Community" whose possibility Dewey discerns in the artful adoption of "the physical machinery of transmission and circulation."[17]

Two qualities unite these disparate uses of "community." First, deep uncertainties about properties and edges: is community about location and face-to-face proximity, or does it consist of affective bonds that can be established by a text message as they are by an embrace? Does it encompass huge swathes of human experience, or is it at best a way to outline a formal arrangement of shared interests? Where is the lower bound—that is, when does a group of atomized individuals, a scattered and manifold accumulation of people and groups, transform into a community? Where is the upper bound—when does a sufficiently large or sufficiently self-reflective community become a "society," a "public," a citizenry, or another communal apotheosis? (And when does a community become a crowd, a mob?) The second quality that binds all these diverse applications of "community" lies in how very nearly impossible it is to use the word negatively, with its many connotations of affection, solidarity, interdependence, mutual aid, consensus, and so on. As Lori Kendall succinctly says, it "carries significant emotional baggage." Raymond Williams summarizes the baggage as its "warmly persuasive" tone—"it seems never to be used unfavourably."[18]

Williams also notes that community seems "never to be given any positive opposing or distinguishing term," though there are many negative oppositions, with new ones being added with each transformation of the word. A whole family of negatives developed when community appeared in its online guise. "Community" was presented as the integument of the "virtual village" produced by "webs of personal relationships in cyberspace," in the words of Howard Rheingold, writing in the 1990s at the very dawn of the web.[19] The atmosphere of free expression online, Rheingold went on, emphasizes the "fragility of communities and their susceptibility to disruption."[20] So susceptible and fragile were the communities, in fact, that "community management" for online groups is a paying occupation with its own evolving best practices and theories today, with a cluster of related occupations such as moderator, community advocate, community evangelist, and social network facilitator. This latest, and in some ways strangest, application of the community concept draws on its history and "warmly persuasive" prestige to manage user behavior and activity on behalf of a given site's objectives, which can mean keeping things on topic, maintaining a civil tone, or strengthening the "brand community" for the sake of marketing—online community, these days, being more often than not a business proposition. (In the 1998 revision to his landmark 1992 hands-on essay "Cyberspace Innkeeping: Building Online Community," John Coate, the astonishingly patient community manager on The WELL, captured the change in the word's use and value from the perspective of the old order: "assigning the mantle of 'community' to one's enterprise before the fact as a marketing hook just serves to cheapen the term.")[21] The disruptions and points of fragility and vulnerability that a community manager works against capture the specific complexities of the idea of community as expressed over infrastructure and code, as issues such as class struggle, anomie, and urban tumult offered points of departure for previous conversations. These are phenomena such as flame wars (ferocious and rapidly escalating arguments), trolling and griefing (strategic provocation and harassment for maximum chaos), sockpuppetry (a person using multiple pseudonymous accounts to create the illusion of support or bully others into submission), and, of course, spam.

Whether meaning bulk, junk, screeds, annoyance, or offense in content or quantity, "spam" is very nearly the perfect obverse of "community," a negative term in both colloquial and specialized technical use that remains expansive and vague, covering a vast spectrum of technical and social practices with varying motives, incentives, actors, and targets. Both words have a productive blurriness that makes them into platforms for development and delineation—for individuals or collectives, markets or nonmarket values, appropriate and just ways to live. Whereas "community" stands in for our capacity to join one another, share our efforts, sympathize, and so on, "spam" acts as an ever-growing monument to the most mundane human failings: gullibility, technical incompetence, lust and the sad anxieties of male potency, vanity and greed for the pettiest stakes—the ruin of the commons for the benefit of the few. We go to community to discuss how people are generous, empathetic, and gregarious—and to spam to discuss how they are suckers, criminals, exploiters, and fools.

As the negative version of the concept of community on the early networks, spam was a significant force in that history as a limit test and provocation. It was a form of failure that helped to define the meaning of success, and it magnified the contradictions that lay within early online communities. It germinated in its many different forms wherever the attention of a group was gathered, whether as a discussion board, a Usenet group, an email thread, or later and more diffuse areas of collective attention such as the results aggregated by Google's PageRank algorithm. With such diverse expressions and agendas carried in one capacious sack, the virtual community demands conversation, debate, and clarification, concerning both the new order produced by its operation and the old order it interferes with or obsolesces. So does spam, interfering with the interference and exposing contradictions within it. In Alexander Galloway's phrase, the rise of the mediating network and its communities obliges us to find new "logic of organization"—because if we do not, the shear between our old models and the new forms will become steadily greater and harder to bear.[22] This shear can be easily seen in most of the domains shaped by information scarcity or secrecy, from journalism, publishing, and international diplomacy to the selective privacy of everyday life that Helen Nissenbaum terms "contextual integrity," in which our friends, family, and professional domains operate in distinct contextual partitions and conversations.[23] Spam presents us with a vital and current—if negative—case of this new logic of organization in action, and it redefines our understanding of "community" online: how it works, and the paradoxes of that work.

The cardinal problem within the virtual community, the problem that spam exploits and aggravates, is the tension between infrastructure and expression, or capacities and desires. One manifestation of this tension—specific to an unusual group, and distinct from the more generic "community," but a readily intelligible example—is what Christopher Kelty terms a "recursive public." This is a public "vitally concerned with the material and practical maintenance and modification of the technical, legal, practical, and conceptual means of its own existence as a public," whose existence "is only possible through discursive and technical reference to the means of creating this public."[24] What makes this example somewhat unusual is that Kelty is describing the culture of open source programmers: people for whom methods of talking and collaborating are something they can easily modify and transform and for whom such transformation is, in fact, a major part of discourse. They operate, first and foremost, from a position of reflexive self-awareness of the means and purposes through and for which they work. By contrast, many of the examples of spam's provocation show us groups of people who have recursion thrust upon them. Akin to Dewey's model of the "public," which is called "into existence having a common interest," with its existence consisting primarily in the ability to "locate and identify" itself, to cohere and mass attention, votes, and money against a perceived negative consequence[25]—spam provides us with *reactive publics.* Obliged, suddenly, to be aware of the means of their own existence and to create deliberate mechanisms that blur between technical, social, political, and legal, these reactive publics must manage themselves and their infrastructure. On the way, they must ask and answer major questions: in whose name? By whose standards? By what methods?

Yes, you may have a "community," with all the emotional baggage that term entails in its dense interlace of shared interest and solidarity, but your community is also a particular arrangement of hardware and software. Your community needs electricity. It is rack-mounted servers, Apache, and forum software, perhaps funded by advertising revenue, volunteers, or corporate largess. (In the case of The WELL, for instance, it was that temperamental DEC computer and six modems in a leaking, poorly insulated office by the sea, a "community" that was always crashing or threatening to overheat.) Your community may be someone else's property and subject to someone else's laws. Perhaps, like GeoCities—or Imeem, Lively, AOL Hometown, OiNK, and so on, in the necrology of dead user-generated community platforms—your community will one day disappear with little or no warning, user-generated content and all. Until it evaporates like a mirage due to a change in business plan, how is your community to police and maintain itself, and how are the rules to be decided? Internet governance is the space of the *really different* (to take Lee A. Bygrave and Jon Bing's term and emphasis). Community on the network is "an artifact that took shape as competing groups struggled in a new technological arena," where the properties of the network dramatically change what is transacted on it.[26]

These network properties themselves can change as well, at different scales and populations of machines and users. Spam's appearance often demands responses on behalf of "us," where "us" can be anything from "a few hundred people on the network we run ourselves," to a vague polity of users on systems hosting millions of people around the world, to "Internet-using citizens of the United States." These different scales create different possibilities for organization, complaint, redress, and the persuasive invocation of "community." (Looking at the history suggests a physical comparison: we can draw distinctions between the quantum scale, the atomic scale, and the galactic scale, because they function under very different types of laws—the elegant simplicity of Newtonian physics breaks down at the edge of the atomic and subatomic, where the strangeness of quantum mechanics takes over, and is subsumed at the upper bound by the still-greater elegance of the cosmic scale, where we can set aside things like chemistry and electromagnetism and work from gravity and mass alone.) The uneasy balance between the group and the means of their existence as a group obtains at every scale we will see through the history of these diverse networks, but this balance and its modification take very different forms in small professional organizations, massive public systems, and within and across the borders of countries.

Spammers crudely articulate this swaying balance between infrastructural arrangements and the concept of a community by exploiting it relentlessly. They work in the space where we are obliged to reflect on our technologies because they both underlie and diverge from our understanding and use of them. This tension is spam's native environment. It is what distinguishes it from other forms of computer crime and why it plays such a role in the explication of communities virtual and actual: spammers take the infrastructure of the "good things" and push them to extremes. Spamming is the hypertrophied form of the very technologies and practices that enable the virtual communities that loathe and fight it. This nature is why it is so hard to define, so hard to stop, and so valuable to our understanding of networked digital media and the gatherings they support. It is this fact about spam that makes it *really different*. This difference is apparent at spam's very beginnings, at the dawn of electronic mail.

ROYALISTS, ANARCHISTS, PARLIAMENTARIANS, TECHNOLIBERTARIANS

Here is the size of the world: one color (usually amber or green) and ninety-five printable characters—numbers, the English alphabet in upper- and lowercase, punctuation, and a space—available twenty-two lines at a time. As we say more, what we have just said drops off the screen. In some very early systems, when it leaves the screen it is gone for good, like a medieval sailor going over the cataract at the edge of the Earth. Everything is scarce: memory, screen space, and transmission bandwidth, which can be measured in characters. It can be hard for a contemporary person to understand just how tight the constraints of networked computing once were, but these constraints are important to understand if the genesis of spam on early networks is to make sense.

Consider the seemingly simple problem of editing text. In the summer of 1976, computer scientist Bill Joy wrote a text editor called "vi," named for being the "visual mode" of another editor called "ex," which allowed you to navigate around in a file and alter the text with key commands. This was hugely useful, because if you were working remotely over a modem, the screen would respond so slowly you couldn't simply click around as we do now, cutting and pasting and adding words. Doing so would be an utterly frustrating exercise and an invitation to mistakes. It was easier to issue commands ("d5w"—for "delete five words") and wait to see the results, keeping some degree of precision in the absence of immediate feedback. "The editor was optimized," Joy recalled, "so that you could edit and feel productive when it was painting [updating the screen to reflect your changes] slower than you could think."[27] (He also recalled that the same summer saw the arrival of the hardware for their terminals that allowed them to type in lowercase. "It was really exciting to finally use lowercase." A *physical chip* was needed so you didn't have to work solely in capital letters.) Except for those with special access to dedicated lines and exceptionally powerful machines, every character counted. The closest someone in the current developed world can come to this experience is trying to browse the Internet over a satellite phone connection, and even that is faster than what Bill Joy was working with.

Even with these enormous limitations, however, networked computing was already prone to the powerful experience of suspension of disbelief and what Sherry Turkle captured as the "crucible of contradictory experience." It is the deep bifocal uncertainty of simultaneously being with dear friends, having a conversation, and being on a hardware and software stack transacted over a keyboard and a screen. With a key command or a few lines of code, another person can suddenly turn this infrastructure against you, and your screen fills, slowly, infuriatingly, with SPAM SPAM SPAM SPAM SPAM SPAM SPAM SPAM SPAM.

About that word: Spam, or SPAM™, was first a food, whose associations with British wartime austerity made it a joke in a *Monty Python's Flying Circus* sketch. Because it's a joke whose humor relies on repetition, and because geeks love *Monty Python*, it became a rather tedious running gag in the early culture of networked computers. By the transitive property of being annoying, "spam" then became a word for other kinds of tedious, repetitious, irritating behavior, whether produced by a person or a malfunctioning program.[28] "In those days," begins one of the many similar folk etymologies of slang on early chat systems, ". . . a lot of people who didn't have a clue what to do to create conversation would just type in their favorite song lyrics, or in the case of people at tech schools like [Rensselaer Polytechnic Institute], recite entire *Monty Python* routines verbatim. A particular favorite was the 'spam, spam, spam, spammity spam' one because people could just type it once and just use the up-arrow key to repeat it. Hence, 'spamming' was flooding a chat room with that sort of clutter."[29] In the bandwidth-constrained, text-only space, as you followed exchanges line by line on a monochrome monitor, this was a powerful tool for annoying people. You could push all the rest of the conversation up off the screen, cutting off the other users and dominating that painfully slow connection. (There exist a number of false etymologies, usually based on "backronyms," coming up with words for which the term could have been an acronym, like "Shit Posing As Mail" or, in a delightful instance of Internet cleverness, the Esperanto phrase "SenPete Alsendita Mesâgo," "a message sent without request." Would that the network that unifies other networks were such a haven for the language that unifies other languages, but there's no evidence for it.)

The word "spam" served to identify a way of thinking and doing online that was lazy, indiscriminate, and a waste of the time and attention of others. Collaborative games with far more complex implied and consensual rules, such as MUDs and MOOs (explained next), are where the mechanical gag of duplicating a line over and over came into its own as a social technology. MUD and MOO, these slightly embarrassing acronyms, need to be explained: a Multi-User Dungeon (MUD) is a text-based adventure game that could support many people at once. These very low-bandwidth virtual realities usually took place in a shared conceptual environment based on fantasy novels and role-playing games (hence the "dungeon," as in *Dungeons & Dragons*). You read descriptions of things, places, and characters and typed in commands to engage with these things and talk to others in the game. You could grab gold coins, throw fireballs, get eaten by monsters, and do a lot of real-time textual hanging out with other players in the fantasy world. One thing you could not do is make deeper changes to the world, such as creating new rooms or artifacts, which is where MOOs come in.

The recursive acronym MOO stands for MUD Object-Oriented; an object-oriented programming language in this case enables users to author new things in the world and even change the operation of the world itself. (An "object-oriented" language makes it easy to swap modular components around without doing a lot of time-consuming low-level programming, letting you connect different pieces together rather than making things from scratch.) MOOs could run the same genre of environments as common MUDs, of course, with elves and talking skulls, but the greater possibilities for collaboration meant they tended to embrace larger and more complex social activities, such as distance learning and world building.[30] MUDs hosted a great deal of annoying, basic "spam" in the sense of dumb *Monty Python* emulation, but MOOs, with their explicit play between the game and the interaction between the users of the game, were a microcosm of the larger event of spam played out on a tiny stage.

An exemplary MOO for this capsule study of spam in (and as) a game is the famous LambdaMOO, a shared world that's proved remarkably productive of academic analysis.[31] A madly proliferating collaborative space, its ambience is nicely, briefly captured by Steven Shaviro: "You took a swim in the pool, grabbed a snack in the kitchen, read some books in the library, and fell through a mirror into a dingy old tavern. You entered an alien spacecraft, and fiddled around with the controls. . . . Close to twenty people are packed into this one space. . . . This one jerk keeps whispering dumb pick-up lines and sexual insinuations into your ear. Someone else dunks you under the water, just for fun."[32] All of this is taking place as scrolling lines of text. The torque of inside/outside the game—the discrepancy between the bodily experience of sitting still at your computer and typing, and the textual expression of a social fantasy of limitless resources in endless communicating chambers—is particularly severe here. Spamming takes advantage of the torque, the discrepancy, by creating objects and programs that can blow a hole through the impression of shared textual space ("You pick up the ashtray. / the_fury says":) by using features of the apparatus which makes it possible ("SPAM! SPAM! SPAM! SPAM! SPAM! SPAM! SPAM! SPAM! SPAM! SPAM! SPAM! SPAM! . . . "), like film trapped in the projector's gate burning up from the heat of the lamp, as the screen fills with vacant white light (". . . SPAM! SPAM! SPAM! SPAM! SPAM! SPAM! SPAM! SPAM! SPAM! SPAM! SPAM! SPAM! . . . ").

Yet spamming means more, in the MOO and elsewhere, than just a prank, like an object coded to spam you when you touch it: it is at once an exploit in the system, a specialized form of speech, and a way of acting online and being with others. It acts as a provocation to social definition and line drawing—to self-reflexivity and communal utterance. It forces the question: what, precisely, are we doing here that spam is *not*, such that we need to restrain and punish it? As with search engines, which will develop increasingly sophisticated models of utility and relevance partially in response to spam's gaming of their systems at points of conceptual weakness, spamming demands higher-order debate from the social spaces in which it operates. In the case of MOOs, you cannot simply write a line of code to eject anyone who says "spam" more than once in a row from the server, because "spamming" could take the form of all kinds of statements and activities, some automated and others not. (The actions of everyone in a room in a MOO are visible to everyone else in that room as a line of text, so a spammer could write a simple program that inputs "kyle_m walks into the sliding glass door and falls down," over and over, faster than anyone else can get a word in.) Before you make serious interventions into the rules governing a consensual space, you must be precise about what you are trying to stop, and you have to work with the users of the space to determine what intervention reflects your collective goals.

As Julian Dibbell describes, LambdaMOO developed a fourfold political structure around problems of misbehavior in general and spam in particular.[33] A rough breakdown of these positions closely approximates the four types of self-definition in the larger social history of spam and antispam over decades, the topic of much of the rest of this chapter: royalists, anarchists, technolibertarians, and parliamentarians. (This typology, like all typologies, is a simplification and misses the nuances of some positions; it is still surprising, however, how many of the developments in the social response to spam in many situations clearly fit one of these four categories.) Royalists want the responsibility of dealing with spammers and other bad actors, as well as the means of enforcement and punishment, to remain in the hands of the "wizards," the systems administrators and others whose access to the system grants them special powers over things like accounts, databases, servers, and the deepest layer of code. Anarchists want minimal interference from the "outside" in any form; problems within the game can be handled by the in-game community. "Technolibertarians," Dibbell's coinage in this context, hold that the "timely deployment of defensive software tools" will eliminate the need for wizardly intervention, collective action, or community governance. Governance itself is the goal of the parliamentarians, who want to regulate wizardly powers, community standards, and "mannerly behavior" through the familiar apparatus of votes, ballots, and governing bodies. There are variations on these forms—such as the mocking term "process queens" on The WELL for those devoted to the endless communal hashing-out of issues in a constant process of therapeutic peacemaking[34]—but those four will do for now.

Spam, writes Elizabeth Hess in her guide to life in LambdaMOO, "refers to generating so much text that its sheer quantity is offensive regardless of its content. Spam can be more than just offensive—it can be disabling for another user who has a very slow communications link to the MOO."[35] Note that this definition does not necessarily mean automated, much less commercial, speech. "Spam" could be someone pouring out their heart in a public forum as well as automatic text generation. It is simply too much, variably and personally defined. Take the problem of "Minnie," a LambdaMOO user whose "long, semicoherent screeds," "'useful stuff incredibly encrusted in verbiage and weirdness'"[36]—which were also aggressively political about the arbitration and structure of LambdaMOO—were not automatically generated and posted but were nonetheless offensive by quantity. The petition to "toad Minnie [that is, forcibly and permanently remove her from the game]"[37] focused on "a long history of vindictiveness, paranoia, slander, harassment, lying, and cheating; but especially her compulsive spam."[38] In other words, her character should be erased and her account closed for being too annoying and antisocial and for writing too badly and too much. Given that "spam" in this situation can include such exceptions and edge cases, how is the group to deal with it generally?

Both the royalist and anarchist solutions to spam were fairly simple: either the wizards should intervene out of *noblesse oblige*, censuring or eliminating players as appropriate, or the general LambdaMOO community should take some form of agreed-upon private action, possibly as vigilantes. The parliamentarian solution took the form of petitions like "toad Minnie" mentioned previously, submitted to the complex arrangement of mediators, wizards, ballots, and voting that constituted "Lambdemocracy." Finally, the technolibertarian solution was based on implementing tools like the "@gag" command, with which you could block out the activities (that is, the text of statements and the announcement of actions) of another user, silencing them for you and you alone. As Dibbell describes the technolibertarian position: "The presence of assholes on the system was a technical inevitability, like noise on a phone line, and best dealt with not through repressive social disciplinary mechanisms but through the timely deployment of defensive software tools. Some asshole blasting violent, graphic language at you? Don't whine to the authorities about it—hit the @gag command and said asshole's statements will be blocked from your screen (and only yours). It's simple, it's effective, and it censors no one."[39] (They will not be entirely invisible to you, though: other people, people you have not @gagged, will be talking about the malefactor's activities, and you will see their conversation.) For many, solutions of this type—whether @gag or systemwide limits to the length of utterances—felt profoundly insufficient, even wrong. They marked "a final transfer of power from the community as a whole to the technology that was meant to serve it, and a naive denial of 'the necessarily social and collective nature of human life,'" as Dibbell summarizes the opposing position in LambdaMOO.[40] Spam's multifaceted character brings out a multifaceted response, a struggle to contextualize, define, and understand what is wrong about it and how it should be handled.

This effect, this demand, is massively amplified outside the tiny terrarium of a platform like LambdaMOO, on the temporal, financial, and social scale of the Internet and the networks that prefigured it. Spam operates in the indistinct areas between what a system explicitly offers and what it implicitly affords, between what it is understood to be and what it functionally permits us to do. In the four cases chronicled in this chapter, spam repeatedly provokes the question it raised in LambdaMOO. What are we going to do about it . . . and who are "we"?

THE WIZARDS

Being a wizard—as many systems administrators and hackers outside the world of games dubbed themselves—offers many pleasures, not least among them the promise of a direct correlation between knowledge and power, and the satisfaction of a pure meritocracy. If you can best another wizard, if you can fulfill your magical responsibilities, it's because you studied harder, because you know more, because you went to the trouble of finding the secrets. You spent long nights over the tomes and grimoires learning the esoteric languages, patiently sat at the feet of more powerful wizards, trekked to remote sanctuaries, and cultivated your inborn talent. While lesser souls were drinking in the tavern, you worked deep into the night in what George Chapman, writing about the wizards of another era, called "the court of skill,"[41] a domain lit by the constellations swinging around Polaris and the glow from the monochrome monitor of your computer terminal.

Wizards are all about capability, and it is from capability that their authority derives. Their relationships with more traditionally vetted powers are often troubled. The kings of Tolkien's Middle Earth don't really know what to do with these bearded strangers who show up from the wilderness with their strange agendas and interests and would likely avoid them were it not for their undeniable powers. Undeniable: "Anyone who can do the work is part of the club. Nothing else matters," wrote Neal Stephenson describing "the ancient hacker-versus-suit [that is, wizard-versus-manager] drama."[42] Gandalf does not crush Saruman because he's better at interoffice politics or writing grant applications; he is the greater mage. Jedi knights, another geek-beloved group kept at arm's length by the society they regularly save, have no patience for the interpersonal niceties of power (that's what the "protocol droids" are for). The only way to join their blanket-wearing ranks is to make your own lightsaber and spend a great deal of time doing unpleasant things on remote planets until you can use the Force. Either your lightsaber works or it doesn't, your spell throws a fireball or fizzles, your program runs or fails. We can talk all night, but your code is either going to compile or it ain't. The Order of Wizards in Middle Earth, the Jedi Council, the Linux kernel programmers: the group operates on "rough consensus and running code," an alliance based on knowing how.

There are many reasons why this perspective has such persistence. It is obviously satisfying to a group of very smart people who feel regularly misunderstood and disrespected by the nitwits who are inexplicably in a position to give them orders: they keep the whole operation afloat, but no one makes the wizard into a king—back to the lonely tower in the forest. (In his novel *Anathem*, Stephenson envisions an extreme form of this hierarchy, in which the IT professionals, technologists, and engineers are effectively a society to themselves, split off both from the monastic scientist-humanists and the distracted and confused populace at large.)[43] It has a refreshingly democratic quality, at least in theory. "The free software community rejects the 'priesthood of technology,' which keeps the general public in ignorance of how technology works; we encourage students of any age and situation to read the source code and learn as much as they want to know."[44] If you want to the join the elite, you just need to work and learn, and the tools to do so will be freely provided. In fact, this last point is one of the great splits between wizardly hackers and the actual sorcerers of history. The latter are zealous hoarders of knowledge, creators of secret languages and allegorical codes and other esoterica. The former do things like circulate multigenerational photocopies of the restricted text of John Lions's magisterial *Commentary on UNIX 6th Edition*, a document with the complete source code of the Unix operating system, to get the material out of the closed world of Bell Labs.[45] (This behavior may also explain the often-noted attraction between certain areas of programming and the more extreme forms of libertarianism and Objectivism: they offer a fantasy of the pure meritocracy applied to the rest of the world, with the enormous relief of denying fate—your failure is your fault, your success the product and proof of your worthiness, intelligence, and hard work, and absolutely nothing else.)

IN THE CLEAN ROOM: TRUST AND PROTOCOLS

ELECTRONIC MAIL IS NO DIFFERENT THAN OTHER MEDIA; EACH GENERATES IT'S [sic] OWN SET OF CONVENTIONS AND RULES OF SOCIAL BEHAVIOR. THE ISSUE YOU'RE ARGUING FALLS INTO THIS AREA - IT REQUIRES THE ADOPTION OF SOME GENERALLY ACCEPTED CONVENTIONS AND SOCIAL RULES. THAT'S ALL YOU SHOULD BE PUSHING FOR - NOT THE ADOPTION OF SOME NEW MECHANISM FOR THE MACHINE. YES??

—Ed von Gehren, to MSGGROUP list, April 18, 1977

Just as it easy to forget how constrained computing used to be, it is easy to forget how small the community of the initiates once was. When ARPANET, the inter-network launched to connect all those mainframes together, could be diagrammed on one sheet of paper, each machine on the network had a copy of a single text file. This file, "hosts," mapped names to addresses and available functions so that you could see who had what you were looking for and connect to them to get it:

```
HOST: 128.89.0.45: CHARLEMAGNE.BBN.COM:
SYMBOLICS-3670: LISPM: TCP/FTP,TCP/TELNET,TCP/SMTP,UDP/TIME,UDP/
TFTP,TCP/FINGER,TCP/ SUPDUP,TCP/TIME,UDP/FINGER:

HOST: 128.95.1.45: KRAKATOA.CS.WASHINGTON.EDU:
MICROVAX-II: ULTRIX: TCP/TELNET,TCP/FTP,TCP/SMTP:
```

```
HOST: 128.103.1.45: CAINE.HARVARD.EDU: XEROX-
8010: INTERLISP: TCP/TELNET,TCP/FTP:
```
[46]

and so on. To make changes to this directory, you emailed a request to the Stanford Research Institute, which kept the canonical copy; administrators were expected to periodically check their hosts file against the master and to update it if there were any differences.

It was in this environment that Jon Postel wrote the 706th "Request for Comments" document "On the Junk Mail Problem," in November 1975. Postel was a wizard if ever there was one, a largely self-taught programmer of brilliance with an enormous beard who was once refused entrance to an Air Force plane until he put on shoes rather than the sandals he wore year-round.[47] Postel was the authority in managing the system of domain names and numbers, which made resources on the network findable, and the editor of the Requests for Comment, or RFCs—circulating documents of suggestions, ideas, proposals, and rules in the community of network architects—from 1969 until his death in 1998. These points are worthy of emphasis because they have much to do with how spam eventually came into being. The predecessor networks to the Internet, and many of the basic elements of the network itself, emerged from contexts profoundly alien to the market and the compromises of democratic governance: these systems were built at the confluence of military contracting—effectively, a radically technocratic command economy—and academic noncommercial collegiality. The meeting of the Iron Triangle (the U.S. Department of Defense, the U.S. Congress, and the defense contractors) and the Ivory Tower and its nascent hacker ethos produced a unique environment. That moment of Postel's lack of shoes is emblematic: an autodidactic hacker who got mediocre grades until he found the world of computers and plunged in, driven by enthusiasm and intellectual challenge, wearing his sandals to meet representatives of the "closed world" (to borrow Paul Edwards's phrase) of the military, which had no shareholders or customers to please, needing only to deliver superior avionics technology. These two very different parts of a Venn diagram meet at the computer on the plane that Postel was helping to maintain. Or we could take the history of the RFCs, the preferred form of publication in the discussions on networking, whose informality—an invitation to discussion, not an authoritative *diktat*—was at the core of the computer networking project (as Johnny Ryan plausibly argues, they are part of the Internet's heritage of "perpetual beta," in which projects are working drafts, the starting point for collaboration, rather than a closed and definitive product).[48] The first RFC was typed in a bathroom, so as not to disturb friends sleeping in the house; it's hard to get less formal than that.[49] RFC 3 reads: "There is a tendency to view a written statement as ipso facto authoritative, and we hope to promote the exchange and discussion of considerably less than authoritative ideas."[50] Anyone who has something worthwhile to contribute can get involved—that first RFC was written not by some mandarin of scientific expertise but by a graduate student. Postel was never appointed, never elected; he was simply, in the words of Ira Magaziner, the Clinton administration's Internet policy advisor, "the person they trust."[51]

Trust is the key word. The people contributing to the evolution of the early network all had either security clearances or the accumulated public reputation that accompanies any academic career, and they drew on the deep but distinct cultures of cooperation that evolved within both academia and the military. These cultures are deeply devoted to techniques of internal interoperation—the academics with their systems of citation, shared disciplinary discourse, coauthorship, and "boundary objects" for research collaboration; the military with its elaborate practices of instilled etiquette and hierarchy—and there is a kind of poetic symmetry in their shared project of getting different computer networks to interoperate. From this background, Postel wrote RFC 706, an engineering document discussing a speculative "junk mail" problem and recommended blocking the malfunctioning host if you were receiving "undesired" material—"misbehaving or . . . simply annoying"—until they fixed their problem. You almost certainly knew the person in charge of the malfunctioning host. You had probably met them face to face; you could just call them on a telephone and ask them to take a look at their machine and get it working properly. Problem solved. (As Abbate describes, ARPANET "fixed" a problem with an overload of network traffic early on by all the participants simply agreeing to throttle the amount of data they sent.[52]) In another RFC circulated shortly after, Postel crystallized the culture of trust and collaboration with a famous recommendation for building protocols to make their robustness and interoperability paramount: "In general, an implementation should be conservative in its sending behavior, and liberal in its receiving behavior."[53] That is, be careful to make sure your host sends well-formed material but accepts all incoming data it can interpret—because the other hosts on the network are very likely producing something of value.

The great microchip fabrication assembly plants that put the silicon in Silicon Valley are built around the production of a strange environment—unique on earth—called the "clean room." The room's cleanliness is defined by the number of particles in the air larger than a micron, a micron being about a hundredth of the width of a human hair. A surgical operating theater has perhaps 20,000 of these particles in a square foot; a clean room has *one*, or none. The air is completely changed every six seconds as nozzles or fans in the ceiling blast purified air down through the grated floors. Objects from the outside world are wiped down with lintless alcohol-soaked cloth; humans are gowned and swaddled to keep them from contaminating the space. The institutional domain of these early networking experiments, the gathering of wizards, was a social clean room where they could experiment and build in a wide-open fashion because the only people on the network were colleagues—and often friends—who knew what they were doing.

Paul Edwards has described Cold War computing culture in terms of a "closed world discourse" built around a fundamental metaphor of *containment*—not simply containing the expansionism of the Soviet Union, but building systems such as SAGE, containing airspace in sensor matrices providing information for a closed loop of feedback-driven, cybernetic strategy, simulation, and game theory. This mesh of language, ideology, and hardware offered a way of managing the huge uncertainties and anxieties of Cold War geopolitics. SAGE, which never worked all that well (a closed world constantly leaking), "was an archetypal closed-world space: enclosed and insulated, containing a world represented abstractly on a screen, rendered manageable, coherent, and rational through digital calculation and control . . . a dream, a myth, a metaphor for total defense, a technology of closed-world discourse."[54] At the deepest roots, computer networking comes from Cold War concerns—Paul Baran, the engineer who developed the fundamentals of packet-switching networks, had begun with the problem of developing a communications system that could continue to operate after a nuclear attack, so the United States would have second-strike capability—but ARPANET and the projects and groups it hooked together were about international interoperability, what we can term an "open network discourse," with the fundamental metaphor being *trust*, for users and hosts alike.[55] If, as Edwards states, the essential Cold War computational vision was that of uncertain and unpredictable forces, from the USSR to the proliferating strategic options of the future, enclosed in a circle of American power, data collection, and information processing, ARPANET's intrinsic image is the network graph of nodes and vertices strung over hastily sketched geographical space, with each node tagged with its particular identity. I say "hastily sketched" because the geography, when it appears at all, is the merest outline, the edge of a landmass—what is significant is the breadth of the nodes: a diverse family of machines, projects, and countries sharing a common protocol. "Membership is not closed," says RFC 3, the manifestation of this idea as a discourse. "Notes may be produced at any site by anybody and included in this series . . . we hope to promote the exchange and discussion of considerably less than authoritative ideas."[56] If the exemplary moment of closed-world computing was the SAGE system, an exemplary moment for open network computing was a hookup in June 1973, when ARPANET was connected by a satellite link to a seismic installation in Norway, and from there by landline to University College London, and from thence to researchers all across the United Kingdom. Make the networks work together, so the people can work together. (With reference to a more recent work by Edwards, we can include this as an instance of "infrastructural globalism," as we can, in a darker and more temporary register, the planet-scale botnets that come later in spam's history.)[57]

All of this development was taking place, in other words, in a bubble of trust and shared understanding, with people vetted by the systems of academic and military advancement and review to make sure they were all "on the same page," on the same master diagram of the network, and sharing a sense of etiquette and appropriate behavior, both on the network and among themselves. After thirty years of rhetoric of the Internet as everywhere nonlocal, disembodied, virtual and cyberspatial, it can be difficult to remember how local, relatively speaking, early networked computing was. It was not the electronic frontier but a fairly small town, populated almost exclusively with very smart townspeople. They had a small-town paper, *ARPANET News* (also available, of course, on the network), whose first issue came out in the spring of 1973. It carried notices of local events ("Theme: Information: The Industry of the 70s?"), touchingly personal notes on meetings ("the atmosphere was purposeful, prosperous, confident"), and "RESOURCE NEWS," amounting to a classified ad section, offering computing capacity and open projects ("Owners of large data bases who would like to store them on the DATACOMPUTER and use its data management facilities should contact . . . ").[58] They were neighbors, in a sense, as well as colleagues, and their collective experience—their experience of collectivity, sharing inclinations and ideas over the well-worn keys of their ASR-33 Teletype keyboards—informed the way they developed protocols and practices. Some months after the first issue of *ARPANET News*, in September 1973, computer scientist Leonard Kleinrock used his ARPANET connection in Los Angeles to get back the electric razor he'd left at a conference in Brighton. He knew his friend Larry Roberts would probably be online (logged in at a terminal in Brighton to a mainframe in Cambridge) and could retrieve it and hand it off to someone going to Los Angeles. He reached across the transcontinental, trans-Atlantic network as though leaning over a fence, shouting across the street.[59]

When many of the core concepts and protocols for electronic mail were forged in the online discussion called MSGGROUP (or MessageGroup or Msg-Group), a conversation conducted over ARPANET, they were built in this social clean room. Steve Walker, starting the discussion "to see if a dialogue can develop over the net," captured exactly this convivial atmosphere: "to develop a sense of what is mandatory, what is nice and what is not desirable in message services."[60] Even if this discussion produced too much material for the taste of a few, they could cope: "mail is easily deleted," wrote Dave Crocker to the group a few messages later, "and so 'junk' mail is not really a serious problem."[61] The actual protocols and the communal mores and norms were being worked out side by side, often in the same document; the rules of conversation and the code for message handling were mutually co-constitutive—design and values, in a loop.

For example, there was the problem of message headers. If we are to have electronic mail, how much information should be in the "envelope" of the message—the part on top, before the body text? There were many different programs for writing, sending, and displaying messages, and they handled messages differently based on their header data. Everyone involved in the discussion was entirely capable of writing his or her own mail systems from scratch to reflect their personal preferences and beliefs. How are messages to be dated? If your mail reader gets a message with the date formatted in a way it doesn't recognize, it may not display it in chronological order, or at all. Messages might be displayed in mangled or difficult-to-read formats as your system tried to cope with a bunch of weird header data from someone else. Should people's names be associated with their computer mailbox address, or should names be portable across the network? What about messages with multiple recipients? These practical issues of standardization led into a still-thornier territory of desires and constraints, expressed in design: if these messages were to be like letters posted in the mail, then they should have a short header, like the text on an envelope—who sent it and to whom, from and to addresses, a date, perhaps a little handling data. If they were to be more like electronic communications, they could carry much more data of value, with a message bearing the route of its own transmission, a picture of the network useful to its technicians and architects, and material about the sender to help you frame the conversation and know how to best reply (if you know what programs someone has available, you can give a better answer to a technical question). Every message could be a kind of encyclopedia about its milieu. Verbose or lean?

Though the word "spam" is nowhere mentioned in their archives (the *Monty Python* troupe had done the skit only a few years before, and there wasn't much of an online populace yet to produce the jargon of popular culture there), they were already wrestling with one of the dimensions of spam. In an environment where attention and bandwidth were scarce, what was the acceptable degree of complexity that people could be allowed to put on the network? Who gets to define what counts as noise, as a waste of resources? Proposals ranged from the simple and minimal, through "redistributed-by" and "special-handling" and "character-count" header elements, all the way to a joke by Bob Chansler that stacks 957 characters of header material (including "Weather: Light rain, fog," "Reason: Did Godzilla need a reason?" and "Machines: M&Ms available but almond machine is empty") on top of a one-sentence message.[62] This struggle was carried out over the MSGGROUP exchanges, in the design of programs, and in the RFCs. RFC 724, "Proposed Official Standard for the Format of ARPA Network Messages," reveals the stunning intricacy behind such a seemingly trivial matter as message headers—and Jon Postel's blistering, point-by-point reply on the MSGGROUP list shows the interlacing of technical design issues and questions of social and political consequence, values, and desires, particularly his reply to section D of the first part of the RFC, "ADOPTION OF THE STANDARD":

> The officialness of standards is always a question at every level of at every level of [sic] protocol. To my knowledge no arpanet protocol at any level has been stamped as official by ARPA. The question of official protocols brings up the question of who are the officials anyway? Why should this collection of computer research organizations take orders from anybody? It is clear that it is in everyones [sic] interest to work together and cooperate to evolve the best system we can.[63]

We can work together; it's in no one's interest to get a bunch of brass and bureaucrats who don't understand the issues involved to start validating every step. "To make a big point of officialness about one step may make it very hard to take the next step," Postel continued. This is a system emerging not from the market's incentives or from the state's command, though both forms of impetus funded and informed it, but through what Yochai Benkler calls "commons-based peer production," with a truly exceptional group of peers sharing ideas, circulating material, and collaborating on improvements.[64] This origin is why such a fantastically esoteric dispute, feuding over header field data, is of real consequence to us. It is an exhibit of the performative ontology, in Andrew Pickering's term, that shaped the production of email and social computing generally.[65] To put it crudely, a group of intelligent peers who trusted each other built systems and standards in which smart, trusted elements operate on the same level. Anyone, machine or human, who can generate properly formatted messages (packets, datagrams, commands, RFCs) can contribute. Bad parts of contributions can be discussed and corrected. Be conservative in what you send, but liberal in what you receive.

An *ontology* is an understanding of the world, a sense of how it is arranged and of the state of things. Blanket generalizations about "human nature" are ontological statements. So are library classification schemes, which arrange sets and subsets in the order of knowledge—*this* is a part of *that* and therefore should be managed in a way that reflects their relationship.[66] To be able to distinguish that a given thing is real and another is imaginary, fake, potential, or virtual is to speak in an ontological mode.

Pickering means something quite complex and nuanced by "performative ontology," with specific reference to his work on British cybernetics in the 1950s and 1960s, but there is one aspect of the idea that has bearing on the discussion here: there are objects, systems, machines, and practices that simultaneously express ontological assertions and show how they could be applied in practice. They make a specific sense out of the world by the way they function and how we function in relation to them. They don't derive from axiomatic claims about the nature of existence, necessarily, but they are, as a practical matter, statements about the way things are, and they stake their functionality on these statements. They "work" as ideas, as well as working as things. The network of networks being built by the people on ARPANET and other computer networks, the eventual framers of the dominant Internet protocols, was an eminently practical system that was also an argument of a kind—for the open-ended capacity for conversation and interoperation between people and between computers and other hardware. We can evolve the standards among ourselves, and the things we produce with those standards will be open to unforeseen participants and uses.

Those unforeseen uses, the immanent social complexity in the system, were already appearing. There were events such as Leonard Kleinrock getting his razor back, legends of a few drug deals done over the portion of the network in Northern California, and much unauthorized conversation about science fiction. And then there was the Vietnam War and the antiwar movement. Prior even to the MSGGROUP, among the wizards at MIT in 1971 there was a case of someone exploiting system privileges, in a way that offers certain precedents to spam, to speak out against the war. MIT was the hub of the CTSS mentioned earlier—the Compatible Time-Sharing System for remote computer access, a separate network from ARPANET. Multiple users on distant terminals—about a thousand in all, both at MIT and other institutions, among them Robert Taylor at the Pentagon with his three-terminal suite—could access a mainframe computer and use it to run programs. Due to the work of Tom Van Vleck and Noel Morris, two MIT programmers, they could also use a form of messaging, a system for forwarding files to particular users, that predated email.[67] Entering "MAIL F1 F2 M1416 2962" (where "F1" and "F2" represent parts of a filename and M1416 and 2926 are specific identifiers—a "problem number" identifying a group and a number identifying a particular programmer) would send a message to Van Vleck, and "MAIL F1 F2 M1416 *" would send a message to everyone on a given project team (in this case, the CTSS programming project itself). For structural reasons, those in the CTSS programming team had a unique privilege: they could type "MAIL F1 F2 * *" and send a message to everyone using the CTSS system at all locations.

"I was mighty displeased one day, probably about 1971," writes Van Vleck, "to discover that one of my team [a sysadmin named Peter Bos] had abused his privilege to send a long antiwar message to every user of CTSS that began THERE IS NO WAY TO PEACE. PEACE IS THE WAY." Van Vleck "pointed out to [Bos] that this was inappropriate and possibly unwelcome, and he said, 'but this is **important!**'" "There is no way to peace" is a quote from A. J. Muste, the Christian pacifist and dedicated anti–Vietnam War activist; 1971, of course, is a period of protests over university-military engagement, two years after the formation of the Union of Concerned Scientists at MIT. Bos used his privilege as a systems administrator to turn this theoretically telephonic one-to-one or one-to-many medium into a broadcast system, one-to-all—and to turn this group of people using a mainframe into the possibility of a politically engaged community of the like-minded. Van Vleck and Morris had created an elegant and prophetic hack for the addressing of files in a set of time-sharing computer accounts, but they had also created an audience heavily weighted toward exactly the group—engineers on defense contracts—that a morally passionate antiwar programmer would want to reach and convince.

This story should be more complex: a new means of communication, a high-minded endeavor, a chilling effect. But the administrator's position over the system, with mastery of the code and the knowledge and the access privileges to change it, was still that of the informal control of the system, creating and banning users and altering the capacities and structure of the network. Of course, these sovereigns are in turn subjected to the authority of the universities, corporations, and governments that employ them—often an uneasy balance of power. As there is a network of networks, so there is a hierarchy of hierarchies in the earliest online gatherings, a problem expressed most succinctly in response to the next protospam message.

INTERRUPTING THE POLYLOGUE

I have never seen any ARPA statement on proper or improper use of the network; I have always just used the "reasonable person" principle. I assume that the military types within ARPA would probably like to restrict access to the network as tightly as they could, and that the academic types would rather that there be free access.

—Brian Reid, to MSGGROUP list, May 3, 1978

What rules actually applied to a statement on ARPANET? We have seen the dispute over how much header data was relevant on a system where every character counted. But everyday life is filled with complex and contextually specific kinds of speech, from gossip to advertising and from political debate to intimate confidences. If the register of ARPANET messages was initially somewhat formal—a conversation centered on how to improve the very tools of conversation—it didn't stay that way. There were antiwar messages, like the one by Bos—though not broadcast so widely and indiscriminately. There were various life-changing event announcements: "I get tons of uninteresting mail, and system announcements about babies born, etc.," wrote Richard Stallman (then addressable as RMS at the MIT-AI computer)—who would become arguably the most important person in the creation of the open source software movement.[68] When Don Woods wanted to get the source code for Will Crowther's computer game Colossal Cave Adventure so that he could improve it, he emailed "crowther" at every host computer on the network, as though he was calling every "Crowther, W." in the phone book. (Woods eventually found Crowther at Xerox's research center in Palo Alto, got the source, and built it out into the text fantasy Adventure, the first of the great narrative-exploration computer games.)[69] Everyone loved to argue. Fast friendships formed among users who had never met face to face—it was an ongoing open-ended "polylogue" in which there was no telling how a topic might evolve, in an atmosphere of informal trust largely among equals, "a select group of people (high school kids regardless!)" carrying on in a strange mix of a high-level seminar, a dinner party, and a correspondence circle.[70]

For some, this was already too much chit-chat, too many unimportant messages to evaluate and delete—and there were more worrisome issues of speech as well. In 1977, a bogus robotics company in New Jersey, Quasar Industries, announced the latest "Domestic Android," a robot that could do various household tasks and "equipped with a full personality

[can] speak and interact in any human situation." This claim was of interest to some of world's foremost authorities on artificial intelligence, who happened to be on ARPANET.[71] It was transparently a twentieth-century Mechanical Turk, and several participants in MSGGROUP began to collaborate on debunking this press conference hoax—"A reporter from *Business Week* magazine is going to Quasar tomorrow morning . . . I want him to take with him a person who would be able to exposte [sic] the thing for what it is."[72] But was it the place of these scientists and engineers on a publicly funded utility to be sleuthing around and using potentially defamatory speech? "We are using US Government facilities to possibly put in a poor light the activities of an 'honest' (and we must assume this) industrial corporation," wrote Einar Stefferud and Dave Farber, often voices of calm reason on the MSGGROUP list. "This could backlash on all of us including ARPA."[73] This statement spawned the full range of responses, from thoughtful agreement to anger at the perceived demand for self-censorship, in a debate that captured the spectrum of ARPANET's self-reflection immediately prior to the first indiscriminately broadcast commercial message on a computer network: the first instance of what would become spam.

That first protospam message was sent to addresses on ARPANET on May 1, 1978, by a marketer named Gary Thuerk and an engineer named Carl Gartley, and it provoked a conversation whose implications continue to resonate—a family dispute among the Olympians that starts to explain why we are at war on the plains of Troy today.[74] The list of 593 addresses for this advertising message included: ENGELBART@SRI-KL, Douglas Engelbart, co-inventor of the computer mouse and key figure in human–computer interaction; POSTEL@USC-ISIB, Jon Postel; FEINLER@SRI-KL, Elizabeth "Jake" Feinler, who ran the organizational Network Information Center (NIC) and, with an executive decision, created the domain name structure of .com, .org, and the rest; and others. (Not all the 593 received it; the addresses, typed in by hand, spilled over into the body text of the message.) The message came on behalf of the DEC, which had been founded by two engineers from Lincoln Laboratory who took the ideas and technologies for interactive computing from the SAGE project and built the iconic Programmed Data Processor, or PDP, series of machines—time-sharing computers beloved by early hackers (the first graphical video game, Spacewar!, was created on a PDP-1 by MIT students). DEC had a strong business presence and established customers on the East Coast of the United States but had much less presence on the Pacific side, and it was Thuerk's idea to take the printed directory of all ARPANET addresses, pick out the ones belonging to West Coast users, and let them know that DEC was having an open house. The message was an ad for the new computers being released by the Digital Equipment Corporation: "DIGITAL WILL BE GIVING A PRODUCT PRESENTATION OF THE NEWEST MEMBERS OF THE DECSYSTEM-20 FAMILY . . . WE INVITE YOU TO COME SEE THE 2020 AND HEAR ABOUT THE DECSYSTEM-20 FAMILY AT THE TWO PRODUCT PRESENTATIONS WE WILL BE GIVING IN CALIFORNIA THIS MONTH." The DECSYSTEM-20 series were the first computers to ship with "FULL ARPANET SUPPORT"; clearly this feature would be significant to the users of ARPANET. They were exactly the people who would want to know.

This message exposed the rift within the concept of "community" on the network. It is the split between community as communication for the sake of shared interests, that is, the community that in its most facile form exists as a market or a target of products, an institutional structure with its sociological roots in the *Gesellschaft*, and the community of "relationships and values," of *Gemeinschaft* (with all the baggage those two categories in turn bring with them).[75] The former articulation of community, a "family" of users to mirror the DECSYSTEM-20 FAMILY of machines, had its defender in the reply to the DECSYSTEM advertising message from Major Raymond Czahor: "THIS WAS A FLAGRANT VIOLATION OF THE USE OF ARPANET AS THE NETWORK IS TO BE USED FOR OFFICIAL U.S. GOVERNMENT BUSINESS ONLY." It was a contractual and industrial arrangement, based on the complementarity of work. ARPANET was not to be used for outside advertising because its attention, bandwidth, and hardware were the property of the institutions that created it for their communication and shared interests.

This argument was not entirely true, of course. There were the personal announcements, the running conversation on science fiction that became the SF-Lovers mailing list, the games: "Who hasn't used net mail for personal communication? Who hasn't spent time playing some new game over the net? Be honest," wrote Debbie Deutsch of BBN in response to the Quasar imbroglio and the suggestion that everyone speak from their official positions as a matter of course.[76] The liberty to write personal correspondence, discuss Arthur C. Clarke novels, and stay up all night playing Adventure hinged on keeping these activities quiet and emphasizing the official mandate. If the DEC ad was inappropriate, why was solving word puzzles in a text adventure a suitable activity? (The SF-Lovers mailing list had in fact been shut down for consuming too much bandwidth; Roger Duffey, the moderator, convinced the ARPA managers, not untruthfully, that it was providing useful practical cases for dealing with large discussion lists.)[77]

A more nuanced position, with the possibility of shared values as well as interests, came from within the user base in Elizabeth Feinler's May 7 message, following Czahor's official statement. She started framing the discussion in her opening disclaimer: "The comments are my own. They do not represent any official message from DCA [Defense Communications Agency] or the NIC." If the network were in fact strictly for government business, everyone involved would simply write from their official position, but there are two networks, and she—a critical administrator for the official network of machines and standards overseen by the Department of Defense—was speaking as a person in the unofficial network, the social graph of the people using the machines. "The official message sent out," Feinler wrote, "asked us ('us' being network users) to address the issue ourselves. I personally think this is reasonable and think we should lend our support or otherwise be saddled with controls that will be a nuisance to everyone involved." The "official message," which Feinler had distributed on Czahor's behalf, is distinct from the group, "ourselves," the "us ('us' being network users)" to which she also belonged. Her import was clear: let's come to an agreement and handle this ourselves so that we can keep our side of the network relatively free of "controls that will be a nuisance." Did the users really want to invite outside authorities in? Feinler's was a parliamentarian position, generating an internal rule structure to mediate between "us" and what she called "the powers-that-be"—to govern ourselves in a compromise with our larger context and prevent further incursions into our space.

This position's character as a compromise was amplified by the response from no less a figure than Richard Stallman, writing without having seen the message yet (still at MIT, he was on an East Coast host and therefore saw only the argument about it on MSGGROUP): "It has just been suggested that we impose someone's standards on us because otherwise he MIGHT do so. . . . I doubt that anyone can successfully force a site from outside to impose censorship, if the people there don't fundamentally agree with the desirability of it."[78] Stallman makes the most basic form of the anarchist argument—"anarchist" as Dibbell means it: not a stand for advertising, spam, or a laissez-faire attitude as such, but for self-regulated standards and values that emerge from the network and are enforced there by the "network users" rather than being imported, imposed, or in dialog with the network's context. This is anarchism in the Kropotkinist mode, in which the "customary law"—our standards, the laws that develop among "the members of the tribe or community"—keep "cordial relations" operating smoothly and functioning best without outside intervention of any kind.[79] Stallman followed up the next day, though, revising his argument with tongue in cheek: "Well, Geoff forwarded me a copy of the DEC message, and I eat my words. I sure would have minded it! Nobody should be allowed to send a message with a header that long, no matter what it is about."[80]

None of these replies quite answered the question raised by John McCarthy, the computer scientist (then at Stanford) who coined the term "artificial intelligence": "The DEC message about the 2020 demonstration was a nuisance. . . . Nevertheless, the announcement was appropriate, and, while the audience was somewhat random, it is probably no more so than the mailing list that brought me a paper copy of the same announcement. Query: leaving questions peculiar to ARPANET aside, how should advertising be handled in electronic mail systems?"[81] Rich Zellich kicked back the idea in circulation at the time that mail users could specify "items of interest" and then would receive "community bulletin board type items" broadcast to all hosts—a kind of keyword-based message filtering far before that approach became a common spam strategy.[82]

"Saying that electronic junk mail is a no-no on the ARPAnet doesn't answer the question," wrote Mark Crispin from the same lab as McCarthy. "I shudder to think about it, but I can envision junk mail being sent to people who implement Dialnet [an experimental protocol for computer systems to connect to each other over ordinary telephone lines, sidestepping ARPANET], and no way it could be prevented or stopped. I guess the ultimate solution is the command in your mail reading subsystem which deletes an unwanted message,"[83] which raises the question of "what ought to be included in the 'self' of self-regulation"—solely the body of individual users, who should take charge of deleting messages they don't want? The protocol developers, who should create a keyword filter in anticipation of the arrival of the marketers and carpetbaggers? Or the ISPs, interested private companies, and national governments, who will build these networks?[84] Who will speak for the polylogue?

Spammers, reliable as rain finding holes in a leaky roof, entered these socially uncertain and definitionally problematic spaces created in the "clean" environment free of commercial speech and its pressures. When the available attention to capture exploded in size, we can see the questions raised here, still unanswered, opening into far broader issues. Spammers began to appear and to become what we now recognize, by operating in the corners created by regulatory arguments where liberty, trust in users, and domains of authority transect unprecedentedly powerful reproduction and transmission technologies.

The antispammers gathered to meet them there.

THE CHARIVARI

COMPLEX PRIMITIVES: THE USENET COMMUNITY, SPAM, AND NEWBIES

People who are buying computers, especially personal computers, just aren't going to take a long time to learn something. They are going to insist on using it awfully quick.

—J. C. R. Licklider, "Some Reflections on Early History"

At the opposite end of the spectrum from the wizards are the newcomers—the newbies, n00bs, greenhorn beginners—with whom every culture of skilled professionals, every craft culture, is forced to find methods of assimilating, indoctrinating, and adjusting. Ted Nelson, when developing the specifications for Xanadu, the original "hypertext" system, presented the "ten-minute rule": "Any system which cannot be well taught to a layman in *ten minutes*, by a tutor in the presence of a responding setup, is too complicated,"[85] which is a convenient rule of thumb for learning an interface—but how long to learn the social rules, the mores and norms, the ways of acting and working appropriate to that environment? Nelson did not have to concern himself with this issue because the complete Xanadu system would have baked all of Nelson's particular social rules about things like quoting, attribution, creativity, intellectual property, and modes of speech into the architecture of the system itself. You wouldn't have to worry much about learning how to behave because there would be very few ways in which one could possibly behave incorrectly in his tightly designed network; the system knew for you. The computer networks that developed out of ARPANET and related projects in universities had no such preestablished social harmonies; they were built in the trustworthy clean room as experiments and platforms for further invention, and they left a lot of vacant space for the development of social rules and criteria for good and bad action. (The struggle over headers marks one of the uncertain zones where code specification ended and social adjudication began.) They left a lot of latitude open to figure out what these systems were suited to—what they were *for.* That open country of adoption and adaptation went double for Usenet.

Usenet was the ARPANET of outsiders. In 1979, as the conversation on MSGGROUP turned to starting a new list, METHICS, that was devoted to working out the code of ethics appropriate to electronic mail, three graduate students began to create the "poor man's ARPANET" for those who wanted to be online but weren't at schools that could land a Department of Defense contract or afford to buy the (very expensive) equipment to join the official network. University computer science departments mostly had massive machines running the Unix operating system, and the students jury-rigged some existing hardware and utilities into a network that would find files on one machine and copy them to another over telephone lines using improvised modems. When you logged into your account on the machine in your department, you saw the new files that the computer had copied from the other machines on the network. You could write and post a file—a message—to circulate to them in turn. That new file, saved on one computer, would be copied to another that it connected to, and from there to all the computers connected with it, until everyone on the network had a copy waiting to be read. You could write a follow-up message, echoing the text of the original, or send a private message to the sender, creating threads of conversation. The students expected that their system would perhaps encompass a hundred computers at most and receive a few new files a day about Unix computing. It would be a grove for the conversation of the Unix wizards—but within four years of the students distributing a version of the Usenet code (on tapes, by hand) to other enthusiasts in the summer of 1980, there were a thousand computers on the network. Intending to start a specialist's colloquium, they had accidentally launched a society.

Part of the original vision of Usenet was that it would be limited by the infrastructural constraint of local calling zones. The computers would have to be connected to others close enough that they wouldn't have to pay long distance rates, keeping the membership limited. (The first Usenet connection was established between Duke University and the neighboring University of North Carolina.) A loophole was found: if you made the computer-to-computer calls that circulated files into a normal, and vaguely unspecified, item on a departmental telephone budget, you could underwrite a potentially global network without anyone ever noticing. More remains to be written about the role of graduate student poverty and parsimony in shaping the hacker ethos, which obliges you to repurpose scavenged equipment, invent workarounds, and maximize limited resources and how this ethos shaped the networks built with those resources. The open-ended, conversational character of ARPANET, exemplified by the RFCs, was even more emphatic on Usenet, where early users built their own modems (earning access with what we could call "solder equity") and invented bureaucratic cover for the machine-to-machine exchanges—and worried about excessive, unnecessary data. Rather than local clusters, where the shared interest was implicit in Unix and proximity, Usenet rapidly turned into a global mechanism for message propagation, and people on Usenet in an Australian university didn't have much use for a message from New Jersey about someone selling a dinette set—a message that famously made its way around the world, propagated from machine to machine, a subject of hilarity and anger.[86] As more computers were added to the network, choosing conversations—that is, having subscriptions, or "newsgroups," so that you could sort out the messages you wanted to receive—became increasingly significant. Here we can already see some of the distant outlines of spam as we know it, which was to be invented on Usenet about a decade from the moment a baffled and furious user at a terminal in Melbourne read about some furniture for sale 10,000 miles away.

The anger sprang from the imbalance Usenet created: it cost nothing for someone to post a message at their local machine, from whence it would be circulated to Usenet, but it cost other users something to receive it, in money, in disk space, in opportunity cost, and in attention. From the perspective of, say, Piet Beertema and Teus Hagen at the Center for Mathematics and Computer Science in Amsterdam—the people who actually had to answer for the consequences if anyone noticed those departmental telephone bills—it cost the equivalent of six U.S. dollars a minute to pull the new Usenet articles off a machine in New Hampshire.[87] Their sheer volume could take ages to arrive over a 300 baud modem, and then what were you to do with the messages? There was only so much room for storage, so the administrator was obliged to delete messages (or dump them to an archival tape, the only reason we have access to most of these conversations now), meaning that as the volume grew, you might miss all the new material if you didn't log in for more than a day or two. It would be gone to make way for yet newer material, and you could well end up trying to catch up on a conversation whose incitement you'd missed, so competition for time on the terminals became still fiercer. What were all these messages eating up the resources?

Steve Den Beste wrote in 1982: "I thought [Usenet] was supposed to represent electronic mail and bulletins among a group of professionals with a common interest, thus representing fast communications about important technical topics. Instead it appears to be mutating into electronic graffiti. . . . [I]t is costing us better than $200 a month for 300-baud long distance to copy lists of people's favorite movies, and recipes for goulash, and arguments about metaphysics and so on. Is this really appropriate to this type of system?" The polylogue on ARPANET had included science fiction talk, ethics, and debates about the Quasar robot, but always under the shadow of the official remit of the Department of Defense, and shaped by the personalities of the professionals who were constructing it. Usenet had no such restrictions (what will graduate students *not* talk about?). "Trivia questions about Superman and Star Trek, and how to use Stravinsky's RITE OF SPRING to retaliate against disco loving neighbors, and why cooking with animal fat will kill you young," Den Beste summarized, going on to advocate that people with such nonprofessional chitchat take it to the burgeoning network of BBSes, which were "run and maintained by private individuals, so any subject is open game." On Usenet, though, where the bills are paid by universities and corporations, "one would think that a little bit of professionalism would be appropriate."[88] What does professionalism mean on this new medium? And what if, rather than being merely unprofessional, the people on Usenet who weren't engaged in behavior appropriate for a university setting were actually inventing a new kind of discourse and a new form of sociality?

Bryan Pfaffenberger has already chronicled the extraordinary struggle between different kinds of actors over what Usenet was to be and where its values lay.[89] The internal friction of "users of the network" versus "the powers-that-be" who owned the machinery that we saw on ARPANET was here amplified dramatically into an out-and-out war for control, complete with threats, sabotage, and a scorched-earth future for the network if compromise could not be achieved. You could create a new newsgroup on your computer—a new flow of topic-specific messages, identified with a period-separated hierarchical name such as "net.chess," for discussing your favorite endgame problems—but that did not mean that any of the other system administrators were obliged to carry it (that is, to copy messages in that newsgroup automatically from your computer to theirs). A person in charge of one of the major sites, from which many others copied their news, could effectively censor a newsgroup for everyone who depended on that site, if he or she desired. To the existing wizardly self-descriptions was added the moniker of "baron," for the powerful administrators whom the conversation had to please if it was going to be distributed, and many of the serfs were not happy with this arrangement. The baron among barons—although he always behaved with pragmatism, humility, and thoughtfulness, he wielded enormous social and technical capital—was Eugene Spafford, who maintained the canonical list of accepted Usenet newsgroups.

Did the Usenet system need to comport itself relative to its semi-official status, supported by the phone budgets and disk space of university CS departments, or did it owe a higher loyalty as a nascent, global, decentralized communications system to the unconstrained free speech and liberty of thought that it made possible? The self-anointed "democrats," fighting quite bitterly with the "feudal" system, certainly thought the latter: "We hold these Truths to be self-evident, that all Humans are created equal, that they are endowed by their creator with certain unalienable Rights, that among these are Unhindered Communications, Unregulated Exchange of Ideas, and Freedom of Speech, that to secure these rights the Usenet is instituted on networks of the world," as Dave Hayes, in full Jeffersonian mode, wrote in his manifesto on the liberties that this new system was ethically bound to provide.[90]

Pfaffenberger sums up the position of the defenders of the most total model of free speech as permitting any speech except that which actively interferes with Usenet's ability to function—that is, that which would restrict the speech of others. Hayes articulates this precisely, showing us the social complexities of the environment in which spam was to flourish by listing his breakdown of what is and is not "net abuse":

Examples of net abuse:

-Posting articles that directly crash the news server that is to inject the post into the news stream.
-Posting articles that contain control messages designed to crash news servers.
-Directly hacking into a news server to disable it.

Examples of things that are NOT net abuse:

-Volumnous [sic] posting
-SPAM
-Excessive crossposting
-Off topic posting
-Flaming or arguing[91]

"SPAM," at this point, was something quite socially specific while still encompassing a whole field of behavior. In fact, everything on the list of "NOT net abuse" except flaming could qualify as "spam" in this period's understanding of the word—posting lots and lots of text ("volumnous"), duplicated across lots of different newsgroups ("crossposting"), which weren't necessarily relevant to the post ("off topic"). Spamming was taking up more than your fair share of that expensive and precious data transmission every night as departments paid to pull in megabytes of data over their modems and consuming the scarce disk space with duplicate text so sysadmins would need to delete the whole batch of messages sooner. Your desire to tell a joke or sound an alarm or spread a chain letter across the largest possible audience would lead you to post the same material across many different newsgroups ("but this is **important!**"). The frustration felt by readers, following the excitement and anticipation of the poll every morning as the new stuff came in, as they noticed the same message again and again taking up room across different newsgroups, was what meant that you had engaged in spamming. (At this point, the offense of spam was distinct from "commercial self-promotion," which had a separate entry on the list of violations of Usenet civility.)

The power of the barons and the "Backbone Cabal," who ran the big sites on whom many others relied, was not needed to keep spam from proliferating, argued the democrats (whom we could identify as some mix of "anarchists" and "parliamentarians" in our typology). There was a lightweight system of social control that constrained no one's liberties and that could keep Usenet in equilibrium. It consisted of public rules (the Frequently Asked Questions [FAQ] lists and guides that came to be called "netiquette") and enforcement through social censure—an inbox full of flames and abuse the morning after your infraction. "Spam" was one of those areas too fuzzy and vague in its meaning to leave to powerful sysadmins to decide. That smacked too much of possible censorship—of the few being able to decide what was useless, trivial, or offensive on behalf of the many.

Here we draw close to the moment of spam's advent in the fusion of political and philosophical anxiety about speech, pragmatic arrangements of technology, and very different social networks competing for control of both the story and the configuration of Usenet. The most baronial and anti-"democratic" of sysadmins ("there will still be the hard reality that for the machines under my control, the current guidelines are simply advisory. I can (and do) ignore certain aspects of the guidelines as I see fit") can be seen as occupying one side of the spectrum of groups trying to shape Usenet.[92] The other edge, the side of an entirely liberated speech, lies still further out than the democrats, however radical, who want a Usenet unconstrained by sysadmin moods—the extreme pole of the debate lay with the anonymizers. The linchpin of the democratic structure of Usenet was a vague general will that expressed itself by angry email and newsgroup posts, making life a misery for bad actors, but this arrangement assumed that the source of the unwanted speech could be found to receive their gout of angry words. Anonymous remailers, such as the infamous "anon.penet.fi," could make your message impossible to trace back to you: you could behave badly on Usenet, and the populace couldn't penalize you because they would have no way of knowing who you were. This was truly free speech, of a sort, as it broke the existing penalties for spamming, and it was troubling enough that a project began specifically to kill everything that came from that anonymizing address.

"It broke loose on the night of March 31, 1993 and proceeded to spam news.admin.policy with something on the order of 200 messages in which it attempted, and failed, to cancel its own messages": such was the ignominious end of Richard Depew's project to build an auto-erasing system to deal with the anonymous message problem, a "solution" that hit news.admin.policy with hundreds of messages under the recursively growing title (here 38 generations in) of "ARMM: Supersedes or Also-Control?"[93] A project designed to manage a threat to the social containment system for spam turned into a disastrous source of automated spam—to the social problems produced by the project of totally free anonymous speech, we can add technical problems produced in putting a stop to it. However, to have the full picture of spam's genesis on Usenet, we must add the problem of money and go back a few years prior to the anonymizer crisis to a pseudonymous scam in the spring of 1988. In its aftermath was an argument about authority and loyalty in a commercializing network and the birth of a new form of collective surveillance and punishment.

Almost ten years to the day after Feinler's message about the DEC computer commercial on ARPANET, on May 27, 1988, a user posted a draft of a letter to the U.S. postal authorities to a Usenet newsgroup and added a comment: "I am concerned, though, whether [sending this letter] is opening up a bigger and possibly more dangerous can of worms than it is worth."[94] The author was responding to an aggressively cross-posted message from one "Jay-Jay"/"JJ"; his response reflects the sense among the Usenet democrats that authority came from organized network users as a coherent and self-declared community, not from the governments in which they were citizens. JJ, the pseudonym used for a small-time charity scam by a grifter named Rob Noha, had rendered this argument more complex. In prior cases of misbehavior, those offended could simply turn to the sovereign power of the relevant sysadmin at the offender's school or business. The administrator could assess the situation, then give the malefactor a lecture, or just kick the offender off the network—while Usenet at large flamed them to a crisp. Noha had posted his begging letter to newsgroups devoted to the Atari gaming platform, to sex, to rock music, to *Star Trek*, to hypertext, to the computer language APL, and on and on. There had been intermittent Usenet chain letters, electronic versions of postal messages propagated by credulous users, but this was something new: a massive simultaneous broadcast across many discussions by an individual directly, in search of money. He posted in waves, sometimes separated by a few days: the 17th, the 23rd, the 24th, and up to the 30th. "Poor College Student needs Your Help!!:-(," he began, outlining his difficulties before he got to the pitch: "I want to ask a favor of every one out here on the net. If each of you would just send me a one dollar bill, I will be able to finish college and go on with my life. . . . If you would like to help a poor boy out, please send $1 (you can of course send more if you want!!:-)" to "Jay-Jay's College Fund," a PO box in Nebraska.[95] The note closed with a prescient request: "PS. Please don't flame me for posting this to so many newsgroups"——a request that reflects either great gall or naïveté.

The crucial detail on Noha's message was the suffix of the email address under which he operated: JJ@cup.portal.com. Portal.com, the Portal Information Network, was one of the first private companies to offer Internet and Usenet access to customers as a subscription business rather than something distributed to students or employees, breaking a key element of the tacit social agreement. (A gradual shift toward moving Usenet news around over Internet connections rather than university computers on telephone lines was one of the factors eroding the control of the major sites and their baronial sysadmins.) Noha's action was in all respects disturbingly connected to the extra-network context of postal systems, currency, and business. Did the sysadmins of Portal owe their loyalty to the rough consensus of "a bottom-up democracy" and the culture of Usenet, or to the business that employed them—itself beholden to investors and customers? Even by the standards of the most vociferous defenders of free speech, Noha's post was tough to defend as a worthy expression.

The parliamentarian reaction to Noha faced some of the same questions of governance and community raised on ARPANET but in a far less academic and trustworthy environment. Who wants to bring in the territorial government? Is there a way we can regulate ourselves? And who will be in charge of those decisions and their enforcement, after so much noise and so many megabytes spent fighting over the role of central authorities versus communal flame attacks? ("Actually, more to the point, does anyone want the FCC or the U.S. Mail snooping around Usenet trying to figure out how to use his postings in court and incidently [sic] whether they shouldn't be exercising more visable [sic] control over such a visable underground communications system as Usenet?")[96] To again use Dibbell's coinage, there was a "technolibertarian" wing that advocated setting aside all this messy social stuff in favor of the "timely deployment of defensive software tools"—you didn't need the Department of Justice or some kind of Usenet star chamber if you had well-developed "killfile" technology to keep you from seeing the messages you didn't want. Many advocated making things so unpleasant for the administrators of Portal.com that they would take the appropriate sovereign action—which, wilting under all the flames, they did, but in an unprecedented manner: "We have received a number of inquiries about JJ. . . . If you view these questions as the burning issues of our time, you might wish to call JJ yourself. You can reach him as: Rob Noha (aka JJ) 402/488-2586."[97] If you want a well-regulated Internet, do it yourselves.

SHAMING AND FLAMING: ANTISPAM, VIGILANTISM, AND THE CHARIVARI

The Portal.com sysadmins' delegation of responsibility was a prophetic act, particularly in what it provoked. The antispam social enforcement that began in earnest with Portal's posting seems like a vigilante movement in many respects: self-organized by volunteers, at times acting in defiance of the law, with the explicit goal of punishing bad actors about whom "there was nothing [the authorities] could do" (to quote Portal's official statement regarding their communication with law enforcement).[98] "Vigilante" is a bad analogical fit in one respect, however, because the enforcers never moved to outright violence. Their methods were prankish, noisy, mocking: collect calls at all hours, "black faxes," ordering pizzas for collect-on-delivery payment, sending postage-due mailings and masses of furious, profane, and abusive email, illegal computer exploits, and harassment of parents, coworkers, and friends of the malefactor. Alleged spammers were surrounded by a constant wasp swarm of threats, trolling, name-calling, and other abuse—almost the mirror image of their violation of the social mores with technically enabled rudeness. Such a response to spam is much closer to the symbolic form of vigilantism called the *charivari*.[99]

"A married couple who had not had a pregnancy after a certain period of time," writes Natalie Zemon Davis in *The Return of Martin Guerre*, "was a perfect target for a charivari. . . . The young men who fenced and boxed with Martin must have darkened their faces, put on women's clothes, and assembled in front of the Guerre house, beating on wine vats, ringing bells, and rattling swords."[100] The charivari was turned against anything the community found unnatural, including marriages between the young and old, widows remarrying during the mourning period, adulteries, and excessive spousal abuse. "With kettles, fire shovels, and tongs," goes a description of a Dutch variant, "the mob hurries towards the culprit's house, before whose door soon resounds a music whose echoes a lifetime does not shake off."[101] "[She] was disturbed by a hubbub in the distance," writes Thomas Hardy, describing a Dorset variation on the charivari. "The numerous lights round the two effigies threw them up into lurid distinctness; it was impossible to mistake the pair for other than the intended victims. . . . [T]he rude music . . . [and] roars of sarcastic laughter went off in ripples, and the trampling died out like the rustle of a spent wind."[102] The "rude music" of banging pots and pans and yelling voices—"discordant voices" is the contemporary meaning of "charivari" in legal parlance—and the march around the house, the sarcastic laughter, the intensely public humiliation and harassment: so the charivari worked in the present case. "Rob Noha / 8511 Sunbeam Lane / Lincoln, NE 68505 / (402) 488-2586 / Phone books are such wonderful things," wrote a user two days after Portal posted Noha's name and number. "Can someone in Lincoln drive by and get the license numbers at his address?" Just as suddenly, the pack disperses, "like the rustle of a spent wind": the charivari is reactive, offering no constructive plan beyond humiliating and shaming the offender, and dies away as quickly as it flares up.[103] (Faced with a more confident antagonist, with the audacious shamelessness of chutzpah, the charivari quickly exhausts its repertoire—as shall be described.)

The charivari form will recur in many different places and different modes, many intimately intertwined with spam, and a bit more deserves to be said about it here at one of the first appearances of the form to frame its history and clearly distinguish it from vigilantism. What is this form, precisely? What I am calling the charivari is a distinct network-mediated social structure, a mode of collective surveillance and punishment for the violation of norms and mores. It evolves from a distinct folkway of early network culture, one to which numerous names have been attached by the recent critical literature—above all the idea of "Internet vigilantes" or "viral vigilantism," which nicely bookend (in 1996 and 2011) the extent of the analysis. David R. Johnson's "Due Process and Cyberjurisdiction" provides the 1996 take, looking at the evolving process of dispute resolution online with its uneasy balance between the sysadmins, who can enforce terms of service, and users, who can migrate to other access providers. "For example, when spamming (sending multiple, intrusive, and off-point messages to newsgroups) became a problem in the Internet, the offended users took direct, vigilante action—flooding the offending party's mailbox with hate mail."[104] Johnson is framing that action in the context of the more effective technical control exerted by sysadmins, such as the ability to cancel accounts and block messages from particular addresses. The wave of flames, the inbox full of messages, is a transitional stage of "summary justice (or self-help vigilante revenge)"[105] that will become less significant than sysadmin methods under the aegis of a developing due process.

Meanwhile, in 2011, as argued by Matthew Fraser, the enforcement of social order by large groups of distributed network users had made a shift from harassing those who acted badly on the network to harassing those who acted badly in life, with mobs gathering by "viral" distribution of videos and reports to shame perpetrators of minor problematic acts—"viral vigilantes."[106] These are, obviously, quite different categories of action, migrating from on-network tools (flaming) to attack on-network individuals (spammers and the like), swamping them with mail, to a mix of on- and off-network tools (news stories and the wide distribution of personal information, telephone calls, pranks) to shame in-world individuals for a variety of in-world offenses. (As explained in the following section, the project of antispam actually encompasses both approaches from a very early moment.) Though as a concept it seems like a natural and intuitive fit, vigilantism is a rather misleading point of comparison, and the charivari concept offers a more nuanced set of comparisons for understanding what's happening in many antispam projects as well as other cases of collective social punishment on the Internet.

Although the idea of "vigilantism" is technically quite diverse, virtually all applications of the term and the idea are connected with violence, and not simply in the sense of hate speech or damage to reputational capital, but in that of people being hanged, shot, beaten, or, at best, run out of a district with the surety of death if they return. Whether we turn to "frontier justice" and lynch mobs, the Klan in the American South, the

Colonial-era "Regulators," the "silent courts" of the Holy Vehm held in the forests of the German Middle Ages, Mormon Avenging Angels, or peasant self-defense bodies like the Big Swords Society of the Qing Dynasty, the particular extralegal nature of the vigilante is that of breaking the state monopoly on violence and basing their authority on the threat of force with the promise of furtherance of justice (situationally defined, obviously; by no reasonable standard could we find the Klan just, but as far as enforcing barbaric community norms goes, they fit the model). Vigilantism has a complex relationship with aristocratic vendettas, peasant revolts, and the mythic power of redemptive and retaliatory violence. It is not only that the rich historical, political, and connotative baggage of the vigilante makes it a bad fit for the present case. It is that a far better fit exists.[107]

The "charivari" described previously is an archaic mode—what a lawyer, M. Dupont, described in an 1832 pamphlet as "conjugal charivari," the archaic folkway for maintaining norms that largely concerns itself with marital misbehavior, whether it takes place in the town of Corneille in the 1830s; Artigat in the sixteenth century; the "butcher's serenade" in London; skimmering, wooseting, or kettling in rural England; or shivaree and whitecapping in the United States and Canada.[108] This action is distinct from *political* charivari, the publicizing of an abuse that sidesteps the legal and governmental process of grievance in favor of a public and viciously satirical and angry display. What we are discussing here is a kind of complex political performance that is built out of mocking laughter, insults, masking and anonymity, and the mingling between active crowds and passive audiences. It is a performance that is at every turn tangled up with the question of who is allowed to participate in political discourse and the extent and role of the law and its representatives. This model does not sound unfamiliar, at this point—for Amy Wiese Forbes, in her analysis of the charivari, it is "broadly the popular politics of satire in public space."[109] Flaming and shaming, from 1988 to 2011, are not extrapolitical vigilante moves but the assertion of a different kind of politics, in which the network's population at large participates directly through jokes, pranks, verbal abuse, public shame, and privacy-violating crowdsourced surveillance, rather than through the mediation of sysadmins, legislators, or police officers.

A few more remarks are needed on the structure of the charivari before we return to 1988 and Usenet gathering to yell at Noha. The charivari, both on- and offline, from the July Monarchy to antispam vitriol and 4chan's lulz-driven crusades in the present day, draws much of its efficacy from renegotiating the boundaries between public and private life. Both the conjugal and political charivari make a public racket around a private home, drawing attention and pressing upon the private citizen the crushing awareness that *everybody knows*—while spreading the word to those who don't know yet. From Noha, and Canter and Siegel (whom you will meet in the next section), and past the turn of the millennium—as you will see in the case of Rodona Garst—the antispam charivari locates the private in the public and puts as much light on it as possible. Inevitably, there is the disclosure of personal phone numbers and home addresses that have been tracked down. Both charivaris are connected with a form of humor—the hilarity and laughter of mockery and lulz—that helps to push the boundaries of acceptable behavior and to lead to a convenient blurring of active crowd and passive audience. On the Paris streets, you could be participating by lighting an effigy afire or by laughing along with everyone else; online, you can join into what looks, in one light, like an attack, simply because you want to participate in the joking—the puns and nicknames and pranks and memes. There is a crucial dimension of pleasure to the antispam charivari that is not trivial. Finally, the charivari, while sometimes delightful and always transgressive (whether involving dogs dressed in ecclesiastical purple driven through the Paris streets, or antispammers circulating and mocking stolen photographs of the spammer Garst) is not ultimately constructive. It may riot, but it does not build beyond periodically establishing areas of public freedom. It is a party, in the sense of catcalls, costumes, and purposeful vulgarity, rather than the sense of a vanguard or political party. And when the fun runs out so does the charivari, without any formal systems left in place for managing the next problem—which brings us back to May 1988.

Activity around JJ/Noha declined over the next few weeks into the boring and confusing aftermath of copycat and counterfeit attempts at replicating the appeal, the whole affair driven off-screen as it went down the discussion thread and further into the past. There had been a brief flurry of activity, charivari glee and skill sharing regarding Noha—many of the messages announcing action, including those forgeries that purported to come from Noha and used a child's don't-do-this reverse psychology to attract more abuse, included their own instructions: "maybe i should put him in my L.sys file with a dummy name, set the time field to Any0300-0400, and queue up some work" (that is, a method to automate calling Noha's number over and over in the early morning hours)—but things simmered down.[110] Usenet returned to trading anecdotes, sharing knowledge, joking, and bickering, the moment of self-reflexive panic slipping into memory. Six years later, as spam began in earnest—in name, in shape, and in response—someone would caution against extreme technical responses, counseling calm in the face of its provocation with the reminder that "we survived Rob Noha."[111] What they faced in 1994, though, was not a mere pseudonymous scammer but a couple of smart, highly public carpetbaggers intent on founding a new order in which they stood to make a great deal of money.

FOR FREE INFORMATION VIA EMAIL

THE YEAR SEPTEMBER NEVER ENDED: FRAMING SPAM'S ADVENT

It was almost exactly six years after the Jay-Jay message and sixteen after the DEC advertisement on ARPANET, on May 11, 1994, that the reminder was posted on Usenet: "we survived Rob Noha." It was implicitly promising that we will also survive this: the first real spam, a commercial, automated, and indiscriminate business proposition, the first message identified as "spamming" contemporaneous with its use—the law firm Canter & Siegel's "Green Card Lottery- Final One?" This was the place where "spam," the word, made its migration from the diverse suite of meanings it had earlier to the meaning it holds now. The old meaning of "spam," a term encompassing the violation of salience—that whatever you were posting, be it duplicated, way too long, saturated with quotes, contextually inappropriate, had broken the implicature of network conversation that held that you should be in some way relevant—simply hopped to the grandest and most ubiquitous and egregious violation of salience yet. Not simply off-topic relative to a newsgroup, it reflected a deeper culture clash—a misunderstanding of the whole point of the network. Noha, in retrospect, became the point at which something should have been done.

There were other, less significant precedents: the "COWABUNGA" attack (a nonsensical, relentless posting across many newsgroups); "Global Alert For All: Jesus is Coming Soon," a classic "but this is **important!**" violation of university admin privileges; the furious autoposting of "Serdar

Argic," who inserted a variable, baffling diatribe about "the Armenian genocide against Turks" into any newsgroup discussion featuring the word "Turkey," whether nation or poultry; and the "MAKE.MONEY.FAST" chain letter. What connected Noha's message and Canter & Siegel's was a thread of money, distinguishing it from the weird verbiage, rants, and piles of duplicate text that had previously been "spam."

The network itself had changed profoundly in the intervening six years—changes in scale, in values, in the meaning of "the network" itself. ARPANET, begun in 1969, had been decommissioned in 1990, the end of the formally academic lineage of networked computing. (The academic research and military wings of the network had parted ways in 1983, when the latter split off into MILNET.[112]) The National Science Foundation (NSF) had created a high-speed backbone system, running the Internet protocol suite, that linked regional networks together, launched in 1988 as NSFNET: a prodigiously growing venue for adoption of the Internet for supercomputers and individual PCs alike. Traffic on the backbone doubled every seven months, and many of the new users weren't computer scientists or programmers. The demographics of what was now becoming the dominant Internet were changing, with NSFNET and the proliferation of personal computers outside office parks and academic labs accelerating a shift in the kind people spending time online, exemplified by Usenet's negotiations and by a little network called The WELL.

The crises of control within Usenet discussed earlier had blossomed into a full-blown social and technical transition. The problem of who got to decide what was bad speech, spam, and egregious self-promotion led to a project that simply sidestepped the barons and the Backbone Cabal in two directions. First, a new news hierarchy was built that started from people's personal computers and avoided the canonical lists of "acceptable newsgroups"—and which spread like wildfire, thanks, among other things, to its inclusion of groups such as .sex, .drugs, and .gourmand. This new hierarchy combined with the move from the initial model of Usenet, in which big computers dial into other big computers and a few major universities are pipelines to many others (called UUCP, for UNIX-to-UNIX Copy Protocol), to a system that could use the Internet instead and thus didn't need to please the sensibilities of a handful of sysadmins (called NNTP, for the Network News Transfer Protocol). The bottleneck of the barons was now looking more like the Maginot Line, outflanked by the opening of new spaces of movement as the Usenet user base grew in great leaps, but the complex social problems to which netiquette was a provisional answer continued, and worsened—another chapter in what Pfaffenberger terms "a long process in which contesting groups attempt to mold and shape the technology to suit their ends."[113]

In the Bay Area, The WELL, which germinated out of some loaned software and Stewart Brand's astonishing social clout and interpersonal skills, had been running strong for years. Unlike earlier networks in which the emergence of social behavior and discussions not specific to computer science had been accidental and problematic, The WELL's reason for being was open conversation among its population of hippies, hackers, futurists, Deadheads, and genial oddballs. It was a system whose ethos sprang not from scientific research collaboration and the formalization of resource sharing between universities but from the "Community Imperative" of communes—"the need to build and maintain relationships between people and to preserve the structure that supported those relationships."[114] Though The WELL remained a small affair, relatively, a business with subscribers numbered in the thousands and a corner bodega's cash flow (especially striking when compared to America Online, which began the same year and soon had millions of subscribers), it attracted a huge amount of journalistic attention and became one of the major models for bringing a diverse population of newbies into the system. Fred Turner perfectly captures both the ambience and the impact of The WELL with the concept of the "network forum," which fuses the "boundary objects" and the "trading zones" of science and technology studies—the ways people can retain the allegiance and useful particularity of their original discipline and community while developing new means of collaboration with others: "Ultimately, the forums themselves often become prototypes of the shared understandings around which they are built."[115] The forums were a classic instance of what Gregory Bateson, one of Brand's mentors, called a "metalogue"—a discussion that is also an example of what's being discussed: in this case, how to share knowledge and have a good time online together.

Although it provided one of the great instances of the value of what would be called "social" in the online services business—users helping and entertaining each other—The WELL also operated under an unusual regime of "community management" closer to ARPANET than life on the open web. To use The WELL, you were logged under a consistent name that others could get to know, in a primarily local network where many of the participants in the Bay Area eventually met face to face, with the constant oversight of moderators deeply committed to using conversation, cajoling, and personal interaction to work out the social kinks. (The sometimes seemingly endless and occasionally exasperating debates over decision-making were known as "the thrash."[116]) It was a group with such powerful social integument that they offer one of the very few instances of *positive spamming*. Notorious, devoted, flame-warring WELL member Tom Mandel started the discussion topic "An Expedition into Nana's Cunt," a loathsome and extended attack on his ex-girlfriend (who was also on The WELL). As the argument about whether to freeze or delete the topic dragged on, other users began bombarding the topic with enormous slabs of text, duplicated protests, nonsense phrases—spam—to dilute Mandel's hateful weirdness in a torrent of lexical noise, rendering it unusable as a venue for his emotional breakdown.[117]

These evolving, struggling networked communities were all living under the looming shadow of a change to the Internet's noncommercial status. While for-profit computer networks existed and thrived, from Portal and The WELL to CompuServe and the leviathan that was America Online (AOL, which is now, decades later, reduced to building an empire of content spam, as discussed in the final chapter), they operated largely on their own islands. The Internet remained noncommercial and was still mostly funded out of institutional subscriptions and government contracts. If you wanted to move data across the NSFNET backbone, you couldn't be making money from it by selling access, and therefore the NSF, not the individual service provider, decided the terms and acceptable behavior for the network's users. There had always been commerce on the local networks, from the BBSes and The Well to CompuServe and others, for people to sell electronics and other goods—recall the infamous New Jersey dinette set—in the community's oversight. As in a medieval town, there was a marketplace in which certain forms of commerce and exchange could happen, and outside those precincts social life was governed by other orders, covenants, and agreements. (Ripping people off happened, essentially in public, and could really be done only once—a method of restricting bad actors that continues, bizarrely, in the workaday world of professional spammers and credit card thieves.) But starting in 1993, the NSF was slowly rewriting the rules by letting money-making

Internet Service Providers (ISPs) run their own networks and the gateway machines that connected them. The question of when and how all these people and their attention and activity was to be monetized grew with the avalanche of early adopters—which brings us to the September that never ended.

Because Usenet was initially based almost exclusively on university campuses—the kind of places that had big Unix machines sitting around—every September a wave of freshmen and new grad students were added to university rolls and started hanging out in the computer lab, and these newbies had to be educated in the rules, netiquette, and norms of Usenet. In March 1994, AOL enabled what it called the "Usenet feature." Its massive subscriber base, which had been functioning inside the enclosed space of their proprietary network, were abruptly turned loose onto Usenet. America Online gave no indication that Usenet was anything other than another one of their properties to which their terms of service and user expectations applied. These new users weren't in any way bound by the old dispensation of sysadmins and the common ground of university affiliation and computer savvy. The grand process of Internetworking—the bridging of all the different domains and their different protocols and populations into the Internet—sometimes resembled nothing less than the abrupt introduction of species from different ecosystems into the same space, like ships discharging bilge and ballast water from one ocean into another and bringing families of organisms, which had been developing into their own intricate coevolutionary niches, into startling proximity. From the perspective of long-term Usenet denizens, who had been patiently hammering out rules of behavior and public order and liberty for more than a decade, it was the beginning of a steady invasion from the vast galaxy of newbies to what was then still a relatively stable little planet on which even the constant conflicts were part of a shared lexicon of debate and a longstanding set of social tensions. It was the beginning of a constant state of siege, an "eternal September" soon memorialized by, among others, the "alt.aol-sucks" newsgroup. Wendy Grossman summarizes the typical experience of two antipodal oceans meeting: "AOLers would post hello messages, old-timers would follow up with vituperative diatribes about reading the FAQ without telling them how to get it, and other old-timers would pile in and take up more bandwidth and create worse useless noise than the AOLers' messages did in the first place."[118]

Two imminent events, yet to come at the moment of the green card lottery, need to be mentioned so that the whole charged environment is assembled around the birth of spam as we now know it. The conclusive and total commercialization of the Internet's infrastructure and the broad adoption of the web both served to open the network up to a new population that was far larger than any it had known before. The first event took place on New Year's Day of 1995. The ban on commercial activity on NSFNET was rescinded, with the Internet ceasing to be the property of the U.S. government as it had been since the first message was sent from Palo Alto to UCLA in 1969. The second was more gradual. Bear in mind that all of the discussions, arguments, and dramas thus far were taking place in the largely all-text and often intimidatingly technical environments of Usenet, Internet hosts, BBSes, and so on. Tim Berners-Lee, at his workstation at the world's largest particle physics research lab in the early 1990s, was working on tools for displaying pages on the Internet that could be marked up with pictures and typesetting and clickable links to other pages. It didn't get much attention at first, but by 1993 the Marc Andreessen–led Mosaic web browser had been built and Berners-Lee had founded the World Wide Web Consortium (W3C). Networked computing was about to become far easier to use for many more people, a vast influx entirely free of preexisting intellectual commitments to the ethos of computational resource sharing, research, noncommercial use, and radically free speech.

This was the distribution of forces in May 1994: the movement away from computer science professionals to the general population, the end of the noncommercial order and the consequent shift in power from sysadmins to lawyers and entrepreneurs as social arbiters, the clash between different sets of established values of use and discourse on different networks suddenly conjoined, and the broad awareness of networked computing in the population of the developed world already expressing itself as growth rapid and enormous enough to create a wholly different kind of space. A phase transition was underway, and in this poised and expectant moment, people logged into their Usenet accounts on the morning of April 12, 1994.

THIS VULNERABLE MEDIUM: THE GREEN CARD LOTTERY

It is important to understand that the Cyberspace community is not a community at all.

—Lawrence Canter and Martha Siegel, *How to Make a Fortune on the Information Superhighway*

In the early 1990s, in response to the lack of diversity in nationalities granted green cards, U.S. immigration policy added a lottery. To enter, you merely needed to send a postcard to the right address, but there was a lot of money to be made for the less scrupulous lawyers in the immigration business by presenting themselves as necessary middlemen who could help with the "paperwork" of registering for the lottery. Laurence Canter and Martha Siegel, a married pair of immigration lawyers in Arizona, were well poised to take advantage of the opportunity offered by the lottery. They needed an advertising platform that was "pretty much the domain of techies and people in academia . . . [who] happened to be foreign born" and therefore interested in U.S. jobs and green cards, and Canter "had been a longtime user of, what we'll call the precursor to the Internet, the online services such as CompuServe."[119]

Their preparations to take advantage of this new possible audience showed both an awareness of what they were doing and a naïveté about the charivari they were soon to unleash. They had tried "a smaller posting [on Usenet] prior to the one that made all the news"—in fact, a couple of them—and had experienced the wizardly repercussions of an angry service provider, so they spoke to their ISP, Internet Direct, about the demands they were about to place on the system. "Our concern initially was that their servers would not be able to handle it because we knew there was going to be huge amounts of traffic generated—both from people's interests and people flaming. And they were really eager to tell us that their servers could handle it, no problem." A programmer in Phoenix wrote a very basic program (a script) for them that would take a given message and post it to every newsgroup they wanted—in this case, very nearly all the newsgroups on Usenet. "I remember executing this script," Canter said in an interview. "I think it was towards the evening when we were doing it—and just watching it run through its cycle. . . . It took maybe an hour or two to post to all of the existing newsgroups at the time, and when it was done, I remember just kind of wondering what was going to happen next."[120]

On April 12, 1994, users of roughly 6,000 active newsgroups logged on to find a 34-line message titled "Green Card Lottery- Final One?" from "nike@indirect.com (Laurence Canter)." "Green Card Lottery 1994 May Be The Last One! / THE DEADLINE HAS BEEN ANNOUNCED," it began, going on to briefly explain the lottery, the ineligible countries, and the "STRICT JUNE DEADLINE. THE TIME TO START IS NOW!!" It ended with all of their contact information, including mailing address and telephone and fax numbers—a detail as telling as Noha's ".com" domain address. Rob Noha, who knew he was trying to pull a small charity scam, used an assumed name and a PO box. Canter and Siegel's inclusion of their contact information made clear that what they were doing on Usenet was legitimate, even normal. It was marketing: of course they wanted potential clients to be able to reach them. (In this transition, we can see the turn that Charles Stivale discerned in early spamming practices on Usenet, the "escalating mores" that go from "playful" to "ambiguous" to "pernicious," as the pranks and conversational domination move into a different zone with far more problematic intentions and consequences.)[121]

It also suggests that for all their sophistication, they had vastly underestimated the scale of their provocation. For a few hours, there was a precious confusion—a relic of the last moment before spam, and advertising generally, became omnipresent online—as people tried to parse the announcement as a meaningful newsgroup post ("We get LOTS of stuff over rec.aviation from Europeans and other non-US nationals looking for ways to train and/or work in aviation in the US").[122] Then the discussion turned to confirmation ("I gave Canter & Siegel a call. Believe it or not, they actually did post this and apparently are proud of it") and retaliation.[123] The issue was not simply commercial speech and irrelevant content. It was that these offenses were being perpetrated to an extent that actually rattled the infrastructural foundations of Usenet and then defended in tones of outspoken flag-draped righteousness quite distinct from the clueless shame or slinking pseudonymity common to prior infractions. Canter and Siegel's script had generated a unique copy of their message for each newsgroup, rather than "cross-posting," which marks the message as a duplicate so that someone who's seen it in one newsgroup can mark it to be left out of others that are to be polled. Usenet as a whole still operated within fairly tight limits of bandwidth, memory, and cost. Two individuals in Arizona had just enormously overconsumed the pool of common resources.

On newsgroups such as news.admin.policy, where much of the ad hoc governance and response of Usenet was worked out, the four political tendencies of anarchists, monarchists, parliamentarians, and technolibertarians came up as before, but some got much less of a hearing. "People always seem to choose this solution since they are prepared to put up with the costs involved," a user wrote in response to the default technolibertarian filtering suggestion, the @gag-ignore-delete-killfile approach. "I for one am not. I don't appreciate getting _the same_ message in every single group I subscribe to. Kill files [intercepting and deleting the messages you don't want] are for pacifists."[124] The parliamentarian approach circled around form letters to the Board of Professional Responsibility in the state of Tennessee, with which Canter and Siegel were affiliated despite being based in Arizona. These letters all had to tread an awkward line, however: "While technically not illegal, this is a violation of the usage guidelines for the network that is offensive, unethical and should be censured."[125] The ultimate nature of the Canter & Siegel pitch, an unnecessary application for a fee, was unethical, but they had not broken any laws with their mass Usenet posting, and the numerous codes of conduct and netiquette FAQ files maintained online in the wizardly/anarchist mode did not add up to a Hippocratic Oath outside the network. The parliamentarians could not bring immediate results by calling on the law and were almost immediately drowned out by the banging pots and overheating fax machines of the charivari—aiming, as before, at making life a misery for both the offenders and the wizards at their service provider.

"Let's bomb 'em with huge, useless GIF files, each of us sending them several, so as to overwhelm their mailbox and hopefully get these assholes' account canceled by their sysadmins," wrote a user on alt.cyberpunk, about a day after the green card announcement (a very early instance of the distributed denial of service attack—a technique for overloading servers that plays a major role later in this history).[126] By that point the Canter & Siegel account on Internet Direct was already overwhelmed. The ISP's service to all its users slowed to a crawl and often went down entirely as the mail server crashed repeatedly under the traffic of complaints, hate mail, flames, and indeed, large files sent to swamp the system. Canter later estimated that "there were 25,000 to 50,000 emails that never got to us" cached on a hard drive as Internet Direct canceled their service and tried to do damage control.[127] The sysadmins were also receiving a very high volume of flames; their email addresses were often listed along Canter and Siegel's in posts for those wishing to make complaints.

Within twenty-four hours their account was closed and any email sent at that point would not do any good beyond further hampering Internet Direct.[128] "How about taking some tape and about three pages of black construction paper" began a user after this news, giving instructions for burning out a fax machine.[129] Reports of phone calls to Canter & Siegel's office were numerous, including those of the "Phantom Phone Beeper," a phone phreak's automated project that called the law office forty times a night to fill the voicemail account with noise.[130] The collaborative research activity taking place on lists including news.admin.policy and alt.culture. Internet assembled a list of email addresses that might belong to Canter, one of which appeared to be for his personal computer. "Is this *their own* private system? . . . Someone please post crack their root and do a kill -9 1"—that is, crash the machine in such a way that no record is left of the state of the crash to help in the recovery, no body for the autopsy.[131]

Their phones, fax line, and computer connections were repeatedly taken out of service, and they were denounced in hundreds of thousands of words, but Canter and Siegel had the one thing truly difficult for a charivari to overcome (as opposed to a vigilante group, with its access to force). They were shameless, possessed of an apparently limitless *chutzpah.* They came back strong, claiming thousands of positive responses and $100,000 in new business for their firm and presenting themselves as icons of free speech on Usenet and the Internet generally, with a total imperviousness to animosity and humiliation.[132] They had "spammed" the entirety of Usenet, as the term was now inaugurated in its new meaning, but they were not vaporized by righteous thunderbolts from the vast and furious community or taken away in handcuffs with their assets frozen. They were giving interviews to the *New York Times* while Usenet's wizards and old hands grumbled on the news.admin.policy newsgroup and made prank calls to their office. "Freedom of speech has become a cause for us. I continue to be personally appalled at the disrespect for freedom of speech by this handful of individuals who would take over the net if they could," as Siegel put it, expressing their basic message from which they seldom strayed.[133] It ran as follows: the Internet was embedded in the territory of the United States of America. There were no restrictions on activity on the network of any real (legal) content. This medium, built largely with public funds, was governed by cliques of weird, ferocious nerds only too happy to dictate when and how outsiders could speak, but they had no power beyond their internal consensus.

This interaction was the crystallization of the other cultural transitions already at work in the end of the noncommercial dispensation, and the arrival of users from outside academia and technological subcultures—people without the shared tacit and explicit understandings that made the whole affair work as well as it did. Two lawyers, not programmers or engineers, turned directly to the context in which the network existed, the context of advertising and business and markets, and above all, government and its laws, and asserted that context against any claims to internal legitimacy that the social structures of existing network users could make. "Cyberspace needs you," they argued to their audience of online advertising magnates *manqué* (with infinite gall, from the perspective of Usenet's long-term users). "Like the Old West with which analogies are often drawn, Cyberspace is going to take some taming before it is a completely fit place for people like you and me to spend time."[134]

As Chris Werry has described in a penetrating study of their spin-off book, *How to Make a Fortune on the Information Superhighway*, Canter and Siegel make a fundamentally Lockean argument (specifically, from the notorious sections on property in the *Second Treatise*), not only for their actions but on behalf of all those who would like to follow in their tracks.[135] The land, the territory, happens to be occupied by "Internet natives"—a very deliberate choice of words, framed in a larger discourse of Wild West "pioneers." However, it belongs to those who can work it profitably. For Locke, the native peoples who were present when the settlers, colonists, and missionaries arrived had no contract structure and no concepts of private property compatible with those of the newcomers and were not engaged in the development of the land and its resources and therefore had no meaningful rights to it. In much the same way, the "natives" of the network, with their strange reputational, volunteerist gift economy, driven neither by state nor by market but by forms of Benkler's "commons-based peer production," had no grounds to be owners of anything. Their eccentric codes of conduct were not merely irrelevant but illegal—Canter and Siegel list much of what constituted netiquette at the point of their campaign, arguing that such a collection of restrictions and caveats, recast as rules for speech in daily life, would be overwhelmingly draconian and unconstitutional.

Canter and Siegel built the defense of their actions, and arguments for future advertisers, around two simultaneous and contradictory assertions, in which we see again the *Gemeinschaft*-to-*Gesellschaft* turn. Their book engages in an "almost hysterical" (in Werry's accurate adjective) argument that there was no "community" on the network, no legitimated source of shared values, "rules, regulations and codes of behavior." There is nothing but a rather horrific vision of "individuals and inert messages," an atomized cyberspace built as though under the aegis of Margaret Thatcher's "There is no such thing as society."[136] At the same moment, they see an enormously lively gathering of "users," "consumers," and "readers" waiting to be exploited—that is, to be refashioned as audiences. They are walking a strange line in which the people using the network are simultaneously without authority over its resources and their development, and yet the people and their attention *are* the resource, that is, the matter to be profitably developed. The confusion of metaphors in this cheap business book is profoundly revealing: there is no community, no people with whom you must negotiate, in this vacant space, and yet the space is made of people and their time and attention, ready to be captured. As Werry documents, within a few years much of the discourse of online marketing was now about "fostering community" around products (spammers had already moved on, as we will see).

David Joselit's *Feedback*—a study of television and video as political phenomena in the 1960s and 1970s—offers a parallel for understanding the layers of what was being done here. Following the vagaries of the use of video technology for political activism, Joselit captures the ambiguity of a "community" being formed around a media platform such as video, with its complex interplay of production and spectatorship: "video and video activism not only deliver audiences . . . but delivers particular audiences by fostering the self-conscious if also market-driven identification as communities or constituencies."[137] Video has the capacity to create new, active spectatorships, fashioning self-aware gatherings capable of action, such as radical feminists or coherent ethnic groups, in the act of taping and seeing an identity formed which one can join. As the corollary to this capability, though, video has the ability to create markets for narrowcasting and niche marketing. The power of video to incite and organize is just that: the power to incite and organize, whether toward projects of political action or purchases and brand awareness. People working together on networked computers was hugely productive of "community," in both of Joselit's senses of the term, and Canter and Siegel saw the latter meaning: gatherings waiting to be tilted, just slightly, to turn into market segments. Peter Bos had seen not just a time-sharing system for coordinating teams and resources but an audience of engineers complicit with the Vietnam War. Thuerk had seen not just a military network for official scientists but the exact demographic for his products. Noha knew Usenet was not just a network for asynchronous, international, noncommercial communication but a huge pool of potentially sympathetic suckers. Canter and Siegel saw readily automated, nearly free, and totally unregulated access to a global market.

The Internet is not video or television, though, and something quite different than a sense of community as a self-reflexive gathering was in danger of being lost. What was under threat was the informational manifestation of that communal sensibility, however problematically formed: the network as a human filtering system for information—as a source of salience. Wendy Grossman, who experienced the moment of the green card lottery firsthand, described the system of Usenet as "Structured so that users have maximum control over what material they choose to look at."[138] Howard Rheingold, in a column written soon after the spam (and published on The WELL, no less) elaborated on this idea in practice:

> The network is valuable because it helps us filter information. A raw flow of unrelated information is no good if you have to sift through a thousand irrelevant items to find the one you need. Computers can organize information, even the kind that accumulates through informal discussions. Newsgroups enable people all over the world to have conversations about topics of mutual interest, and to search those conversations for valuable information . . . the vast but organized conversation makes it possible for you to read only the newsgroups devoted to the topics that interest you.[139]

The etymological transition of the word "spam" was not trivial but spoke to the deep problem and fear that Canter and Siegel had provoked. Although there was friction around commercial activity, people had been selling furniture and Grateful Dead tapes on the network and getting clients for massage or for programming since it had begun to move into the civilian sphere. The difference here was that Canter and Siegel had broken the rule of salience, the rule all spammers always break. It's not that they brought commercial grasping into some utopia of loving-kindness and cybernetic anarchism. It's that they violated the principle of staying on topic and ignored the importance of salience and the skills entailed in knowing how to get what you need. Commercial advertising took the title of "spam" from the bad behavior of flawed inhabitants of the network because it was the *same thing*: material irrelevant to the conversation, violating the implicature out of which meaning was made and wasting attention. From this point—from the question of how to increase the value of the conversation being built out of the data—spring search engines, search engine spam, and the problem of "junk results"; social networks and data filtered by significant relationships, and the crises of social spam; and email spamming, whose slow-brewing disaster begins in chapter 2.

To bring their story to an end, there were penalties related to Canter and Siegel's project in the longer term: Canter was disbarred in 1997 by the Board in Tennessee, partially as result of the green card lottery campaign, and Cybersell failed because their infamy online made Internet service providers increasingly dubious about dealing with them. "Bit-by-bit, yes, we were terminated by pretty much everybody," Canter later said.[140] (The magazine *Wired* refused to carry any advertising for their book.)[141] But at the time, the effect on Usenet was devastating. Canter and Siegel forced the hand of the network, and it did not have much to put up aside from telecommunications attacks, social intimidation, and plaintive appeals for good behavior: "NO COMMERCIAL ADVERTISING, if you please. Imagine what this vulnerable medium would look like if hundreds of thousands of merchants like you put up their free ads like yours."[142] "This vulnerable medium" is an interesting phrase, at once an overt plea for responsibility presented from a position of weakness and a more covert assertion of the need to maintain the status quo in favor of the people who already run it. It recalls the very real and cogent concern for the networked commons but also the all-too-common threat language of groups in power who would like to stay that way—the rhetoric of a vulnerable democracy in a dangerous world used by the Bush administration in its erosion of civil liberties, for example. Problematic statements, complex to make and complex to defend, flaked off of the charivari enforcement model as it fell apart. Something more coherent to fight spam was clearly needed, whether that was to take the form of a social tool, a legal decision, a technical fix—or all three. Already the ads for a fat-loss cream from a PO box in Miami had arrived, and the strategies for using email as well as Usenet for advertising were taking shape. *How to Make a Fortune* was selling, and the disputes around context and community were turning into a very different conversation about money and law. There was no time to lose.

2 MAKE MONEY FAST: 1995–2003

INTRODUCTION: THE FIRST TEN MOVES

The theory of the opening is one of the most fascinating aspects of chess. The options are initially so limited, and the areas of focus still few. From the first move, the complexities proliferate with an enormous multiplication of the space of possible future moves, and long term strategies open like fans as both sides mobilize and threats appear. Within the first ten moves the game is often decided, and the remaining twenty or thirty moves are just the gradual confirmation of the truth. To manage this complexity, chess players rely on a cognitive approach called "chunking," as popularized by Douglas Hofstadter via the work of Adriaan de Groot: "There is a higher-level description of the board than the straightforward 'white pawn on K5, black rook on Q6' type of description, and the master somehow produces such a mental image of the board."[1] They think in terms of blocks of pieces and moves, associated sets of actions without having to sort through all the possible choices (most of which are useless or counterproductive). They have engaged in some "implicit pruning" of the tree of choices.

So far, we have been able to follow spam's precedents roughly chronologically, move by move, across a diverse family of networks: CTSS, ARPANET, Usenet. Now the board comes entirely alive as a dense matrix of overlapping and interacting actors and forces—the infrastructure of network protocols, hardware and standards, legislation and political frameworks, companies small and huge, financial events, activist groups, hackers, lawyers, demography—with feedback loops, arms races, struggle over resources, and reinventions all going into making spam. Here, we need to begin chunking as chess players do, gathering the common threads and significant moments "to assemble," in Latour's phrase, "in a single, visually coherent space, all the entities necessary for a thing to become an object."[2] This total picture will show how the nebulous lexically mysterious act and thing called "spam" gets turned into something that can be specified by law, by terms of service and acceptable use policies, by programmers developing new algorithms, and by interested parties staking claims on how the Internet is to be used. The story of spam was through 1994 largely a story about building community and managing scarce resources on networked computers and in the process defining misbehavior, marking it, and stopping it. From 1994 on, the threads of the concept of community and the capture of attention entwine with issues of money, collective organization, and the law. In the final chapter, which covers from approximately 2003 to the present day, these contentious affairs are both expressed and altered by struggling algorithms. Each of these layers builds on and continues the one before, of course; the history of contesting algorithms isn't complete without understanding the role of laws, and the history of the laws is only half the picture without the intricate dynamics of many conceptual communities that, contrary to Canter and Siegel's claims, were tumultuously alive and growing online.

Consider Sanford Wallace's apology. Wallace is one of spam's more tragic figures, in a pretty dense field—for all the gall and brash chutzpah of the business of spam, its heyday attracted an astonishing number of notably isolated and seemingly damaged people whose shadiness had none of the glamour of the criminal rebel or the technical finesse and daring of the security-penetrating hacker. The work of spamming in this period involved very little physical danger and no threat from the law, but required absorbing an absolutely astonishing amount of verbal abuse and public loathing. The crop of post-1994 spammers who accepted this arrangement included a few high school students, a failing neo-Nazi, an ex-MIT artificial intelligence (AI) graduate student currently in hiding, disbarred lawyers, failed insurance agents, the kind of pill-selling quacks who used to advertise in the back pages of weightlifting magazines, white-collar felons, and get-rich-quick schemers. Many of their stories end badly, with prison time and massive fines (for crimes like fraud, forgery, identity theft, as well as violations of antispam laws), and a few more tragically. It is a troubled population that despite much evident ingenuity and entrepreneurial zeal has a hard time breaking away from the business even as it ruins its leading figures. Wallace is emblematic of this, a sometime DJ and owner of a bankrupt New Hampshire nightclub who embraced his identity as a spam impresario in the 1990s, registering a domain after his nickname "Spamford" and dubbing himself the Spam King (a term applied to several others, and the source of the title of Brian McWilliams's excellent 2004 book about this era of spam; his reporting for *Wired* on spammers remains one of the best sources for the everyday biographical realities of the business during its boom years).[3] The arc of his career includes "junk faxes," which became part of the legal precedent for fighting spam, conventional email spamming, systems for placing ads on people's browsing software, an inventive scheme to provide free Internet access to users willing to accept a certain amount of advertising mail at their accounts, and finally social network spamming on both MySpace and Facebook. He has also faced a string of lawsuits, almost all of which he has lost, including enormous fines—$230 million to be paid to MySpace and $711 million to Facebook—and charges of contempt of court. (The suit by CompuServe against his company Cyber Promotions Inc. is the source of an opinion that, in a truly peculiar footnote, draws on *Monty Python* to describe the etymology of spam.)[4] After each trainwreck of a case, he eventually turns up again, in another iteration of the business.

Yet his career as a spammer ended first in April 1998, when he issued a contrite apology: "I'm backing out of the deal. I will NEVER go back to spamming. . . . I apologize for my past actions."[5] To whom was this regretful statement addressed—a judge, an industry body, an ISP, the general public? No: it was a letter of apology directed to the volunteers in the antispam gathering at news.admin.net-abuse.email (NANAE) on Usenet. "I have a new respect for many of the regulars here. It is now clear to me that most of you *are really here* to stop spam—not just for the thrill ride. . . . You folks are WINNING the war against spam. My fight is over." This ad hoc group *was* the authority at that point, as far as it went; they knew spam better than anyone aside from the spammers themselves, and the two groups had thoroughly charted the legal and social gray areas enabled by the tightly coupled arms race of technological means and tactics.

The story of spam from 1995 to 2003—the rise and fall of spam as a business that could feign respectability and in the process reshape the culture of the web—is the story of the million-plus messages that constitute

NANAE. But it is also the story of doomed "blocklist" projects, which would log the addresses from which spam came so they could be blocked—an alphabet soup of projects with acronymic names like ORBS, RSS, RBL, MAPS, IMRSS, and DUL. It is the story of the travails of other antispam initiatives such as metered email (charging a small amount of money per message so that senders working in the thousands or millions of messages would be priced out of a profit) and opt-out lists so that one could in theory unsubscribe from responsible spam operations. It's a governmental story, too: the story of legislation such as HR 1748, HR 3113, S. 711, S. 875, CAN-SPAM, California's Business & Professions Code §17529.5, and many other state laws (to say nothing of the legal history overseas). It is acronym-laden story of activist antispam organizations such as ROKSO, CAUCE, MAAWG, and ESPC. Of course, it must be a story of software and its circulation, of tools to stop spam (procmail scripts and cancelbots) and tools to produce and send spam (First Class Mail, GeoList Professional, and the ominously named Avalanche). And, of course, it must also be the story of spammers: Alan Ralsky, Leo Kuvayev, Scott Richter, Laura Betterly, Robert Soloway, Sanford Wallace, and many others—all those creators of different scams, brilliant workarounds, cheap tricks, and peculations. It is incomplete if it is not the story of spam going global, with regional variations being developed by West African confidence tricksters, and under-regulated overseas ISPs providing safe havens to spammers. And finally, it must include the story of the emergence of new forms of spam designed to manipulate search engines, which evolve in response. The board is alive, now, and the story has become enormously complex.

This is where our chunking comes in. There are several ways to cut through this history while producing an account comprehensive enough to capture that transition from thing to object—from spam as something unusual, a specific negative part of the texture of a community, to an infuriating but ubiquitous or even normal part of life online. Our approach will follow spam as it becomes entangled with laws, national boundaries, and the evolution of online culture, with the layer of algorithmic mediation growing in influence. This chapter brings together four chunks to provide an overview of the network getting rich and exploitable, getting public, getting complex, and getting truly global—a transformative process of "massification" (to use a term from Matthew Fuller and Andrew Goffey).[6] The first chunk is an exemplary struggle between a spammer and a vigilante hacker, Rodona Garst versus the Man in the Wilderness. The second chunk chronicles the antispam battle on Usenet, with the cancelbot wars and the foundation of NANAE. The third chunk covers the history of "Nigerian 419" messages. Finally, the fourth chunk outlines the coevolutionary dance of search engines and their spammers.[7] Like everything about spam, this study starts with someone seated at a computer: in this case, Rodona Garst, of Premier Services, in the summer of 2000.

THE ENTREPRENEURS

Six years after Canter and Siegel's Lottery message, the Internet in 2000 is profoundly different from Usenet in 1994. Spam has been discussed in newspapers and on television and the Congressional Record. (In fact, spam crashed the U.S. House of Representatives mail servers the year before in a classic communications morass of one mass spamming followed by furious reply-alls.)[8] It is now the subject of numerous laws from California to Austria. The market has diversified. There is, of course, pharmaceutical spam, from the "Skinny Dip" fat-loss cream advertisements that proliferated on Usenet following the lottery message to the archetypal Viagra and herbal supplement messages. There are spam campaigns devoted to mortgages and debt consolidation, to watches and perfume, to laetrile, to "free government grants" and guides and software so the recipient can join the business themselves, and, of course, to pornography. Spam for a Nissan dealership in Seattle has provoked the first picketing of a noncomputer business for online offenses, after the dealership paid $495 to a Las Vegas-based company called The Wizard to message 21,000 purportedly Seattle-based addresses.[9] Spam has become a domain of proliferating niches, serving a thousand contingent and particular lines of work in need of an inexpensive and unregulated way to find an audience.

Consider the problem of advertising phone sex lines. The business of phone sex is structured around arbitraging the different settlement rates—how much it costs to call a given country from the United States. A company in the United States leases lines in another country to route the calls and takes a per-minute cut of the settlement rate, with most phone sex calls routed through places like São Tomé, Moldova, and the Republic of Armenia. These millions of minutes of pay-per-minute activity were a significant source of income for the leasing countries: foreign pay-per-call operations were an enormous part of the traffic on Guyana Telephone and Telegraph (GT&T) circuits, for instance, making up $91 million of GT&T's $131 million of revenues in 1995, and São Tomé kept approximately $500,000 of the $5.2 million worth of phone sex calls Americans made via their country in 1993, using the money to start a new telecom system. It is one of those strange macro/micro moments that will recur on the fringes of spam's history as lonely, sexually frustrated Americans unintentionally built telephone infrastructure for an island they'd never heard of off the coast of central Africa.[10]

This big business had a single bottleneck: it was largely run on a franchise basis, making it doubly difficult to advertise. The "distributor" model meant that you leased a line from the phone sex company, and for the right to advertise it there was usually a one-time fee of between $500 and $2,000. The company handled the technological and administrative overhead involved in cutting deals with the Armenian telephone company, provided the content, and gave you a percentage on the minutes callers spent on your lines (probably about twenty or thirty cents to you for every dollar the company made). How much you make as a distributor, therefore, is entirely contingent on how much you advertise your numbers in a very crowded and competitive field. As it happens, the service provider you've paid a thousand dollars has suggestions, like the back pages of weekly papers, stickers in phone booths, and sending lots and lots of email, or subcontracting with a professional in the business of sending out an enormous volume of email for whatever their clients want—which is how the "distributor model" (letting people in hotel rooms talk with slumming actors and stay-at-home parents in need of a second income) became part of the enormous tide of late-1990s spam.[11] Now multiply that story—that complex scenario of a unique industry with an advertising gap—over and over again.

Spam's boom years led to a rapid speciation of models, which can be roughly grouped on two axes: for yourself or for hire, and legitimate or crooked. Some people sent spam to sell something of their own, whether that be pornography or pills, deadstock shoes, or software promising an entrée to the spamming business. Some were in the business for hire, sending spam on behalf of others. Some framed themselves as honorable and even noble—just like other advertisers but working for smaller clients who couldn't normally get an audience, and defending free speech—or as bold visionaries (leaning heavily on the "cyber-" prefix, in the Canter and Siegel mold), or merely as marketers like any other, and in every case undeserving of the vitriol they received. Others framed themselves as mercenaries, doing whatever needed doing, including scams and brazen porn come-ons. They worked covertly, as grifters always have, with overseas hosting services, pseudonyms, and misleading return addresses. The vast majority fell into an intermediate space on those axes.

Out in the corner described by *legitimate* and *for-hire* is Laura Betterly. An accountant who wanted to work from home with her children and started a "bulk email" business, Betterly claims to have never done a campaign on behalf of adult services or disguised the message source and subject. She speaks at marketing industry events. Her company charged something on the order of $1,000 for the first million messages, less for more, with package deals available and with additional charges for the sales leads they generated.[12] She felt quite comfortable speaking out against the "hate groups that are trying to shut down commercial email."[13] "It's what America was built on," she said in an another of the many interviews she gave in her company's heyday. "Small business owners have a right to direct marketing."[14]

In the opposite corner, *crooked* and *for themselves*, was someone like Davis Hawke (and his protégé Brad Bournival), the failed neo-Nazi whose bizarre career in spam was exhaustively chronicled by Brian McWilliams. He ran huge numbers of throwaway domains such as soothling.com and jesitack.com under the cover of Quicksilver Enterprises and the wonderfully titled Amazing Internet Products, hosted through particularly disreputable overseas service providers. He created an erectile empire on sales of an herbal pill called Pinacle (spelled as such)—those with email addresses in circulation in the early 2000s may recall the pill guaranteed "to add up to 3"!!"—that helped to fund periods spent in hiding somewhere in the northeastern United States.[15]

The two are an ideal study in contrasts. Betterly has a new company (search engine optimization, Google ads, social media marketing) and lives as a marketing professional in Florida. After winning a $12.8 million judgment against Hawke in his absence, AOL recently considered taking a backhoe to property belonging to Hawke's relatives in search of the precious metals he had purchased and may have buried. Hawke is currently in hiding, possibly in Belize or Thailand, the court filings speculate, or lurking somewhere in the cold and rusting backwoods towns near the Canadian border.[16]

An enormous number of spammers worked between these extremes. They operated in the quite broad gray area of nebulous state laws, jurisdictional confusion, the uncertain authority of etiquette and custom left by the prior order of "community" online, and the open rhetorical space created by the Direct Marketing Association (DMA) and other stakeholders who didn't want to close off this potentially fertile space for advertising. ("The government, however good its intentions, should not strangle electronic commerce at birth" was typical of the DMA's statements, as they tried to develop their own systems for regulating spam without making it more difficult for their member organizations to send ads.)[17] People could migrate from one end of the spectrum to the other, as did a young man named Chris DeWolfe with a jumbled sack of business interests in the Los Angeles dot-com scene. DeWolfe ran an email marketing company called ResponseBase, which used the mailing list built up by a previous, unsuccessful company to sell everything from ebooks (with delightfully perennial snake-oil titles such as *How to Date Pretty Girls*) to cheap remote-controlled cars. "During flat or low-growth economic times, direct marketing methods have proven to be the most efficient method of deploying marketing dollars."[18] ResponseBase also sent out mailings on behalf of others, including debt-consolidation services and struggling LA startups. The remarkable success of ResponseBase led to its owners' role (after many transformations) in the creation of the company whose domain DeWolfe had purchased after the original owners went bust—MySpace.com, at one point the world-beating social network. A surprisingly robust business model could be built around selling other would-be spammers software, advice, and addresses rather than spamming on one's own behalf. Jason Heckel sold pamphlets of instructions ("How to Profit from the Internet"); Davis Hawke's *The Spambook*, a partially plagiarized guide to spam success, came bundled with two address-harvesting programs; the pseudonymous "Email America" posted regularly in AOL chatrooms, offering to sell batches of millions of addresses in 1996.[19]

Rodona Garst and her company, Premier Services, were exemplary gray-area spammers. They had a public face of sorts, doing for-hire client work out of an office in Tennessee, with satellite employees elsewhere in the country and a marketing brochure full of vintage dot-com handwaving about "opening the doors to cyber space and smoothing the transition into these infinite markets."[20] They had a diverse spread of campaigns going—for diploma mills, for a "turnkey e-commerce solution" for merchants, for "Growth Hormone" quackery, for pornography and for multi-level marketing schemes ("You are potentially 60-90 days away from never having to 'go to work' again"). Most notably, they did campaigns for yet another new spam business model, which was quite a strong one: "pump-and-dump" stock touting, promoting penny stocks on behalf of brokers to push them up to a manufactured peak just in time to sell and move on. We know all this because Garst, or one of her employees, spoofed the return addresses of spam messages—a fairly common practice that makes the spam message appear to come from someone other than the actual sender, leaving the spoofed address to receive all the angry replies. This time, Garst's group repeatedly used a domain that happened to belong to an ISP's security consultant. He received all the bounced undeliverable messages, and all the flames and other hate mail afterward, plus the threat that his domain might be blacklisted by other ISPs. ("He" remains anonymous, but rather melodramatically dubbed himself the "Man In The Wilderness," hereafter MITW, so we will use the male pronoun.) Our anonymous hacker took it upon himself to infiltrate first Rodona's office system, and then, over time, the company's servers and the machinery belonging to all the remote employees. Then, in a charivarist's gesture that is becoming familiar, after "quite a bit of soul searching" he put about 5 MB worth of data from their computers online. It included chat logs, emails, order forms, screenshots, personal photographs, even a brief audio file generated by Garst testing her computer's microphone. He called this collection "Behind Enemy Lines."

LET'S GET BRUTAL: PREMIER SERVICES AND THE INFRASTRUCTURE OF SPAM

Part of the process of studying spam is learning to think about the kind of objects through which we can study it, particularly the nature of the archives and corpora. The prior chapter of this book was built out of the archival text files in which many of the exchanges on ARPANET and Usenet were stored, backed up to tape at the time, and then copied, consolidated, and uploaded. One significant artifact in the following chapter will be the enormous archive of email internal to the Enron corporation that was released to the public as part of the Justice Department's investigation and then refashioned into a spam-analysis corpus. Each of these and the many other elements that play a part in research need a slightly different and contextually aware approach. The 5 MB of data that the MITW posted are enormously interesting and problematic and require a special kind of attention.

To begin by stating the obvious, the data was obtained illegally. The MITW does not detail his methods exactly, and antispam activists gave voice to a good deal of speculation as to how he had the access to engage in such a thorough sweep of the hard drives of Premier Services. Was he a disgruntled ex-employee, a fellow spammer taking vengeance or eliminating a competitor, or what he in fact purported to be? As a result of all this secrecy, the provenance of the data is deeply uncertain. The MITW claims to have copied more than 100 MB of data from Premier Services, of which this small redacted portion is all that's made available. (He also claims to have found 1,300 AOL account usernames and passwords on Garst's system and passed those along to AOL to encourage them to crack down on Premier Services.) "Redacted" is not quite the right word, however, as the data are full of people's names, postal and email addresses for clients and friends as well as those internal to the company, and even bank account information.

All this information is framed by its presentation as an exposé website, with the very particular aesthetic Olia Lialina labels the "digital vernacular."[21] The text is in Times New Roman, in a variety of colors, foregrounded against a tiled image of one of the classic crushed velvet backdrops beloved of amateur web designers, "texture_blue." The data are contextualized by MITW's narrative and editorial comments. The site was quickly reproduced in multiple, separate locations—a process called "mirroring"—lest Garst have it taken down, but the mirror sites themselves have variations. In particular, after Premier Services ceased operation, the administrator of the most prominent mirror decided that the site was a matter of historical interest: "I've decided to take the 'Lets Get Brutal!' section offline on my mirror."[22]

The so-called Brutal materials include photographs of Garst in various states of undress in the office and at home, as well as what appears to have been a private boudoir photography session for a Premier Services employee and two erotic stories allegedly found on one of the servers. The MITW lists ten reasons justifying the decision to publish them. The list is a catalog of spam outrages ("flooding the Internet with millions of spams every week for five years") and the failure of legal response. The Brutal files are the best, and therefore worst, instance of charivari shaming as an antispam strategy, a deeply invasive and unsettling act of social pillory. When does a criminal forfeit all right to his or her privacy? Especially, it must be said, a "criminal" who has not been convicted of any crime by anyone other than a mysterious hacker-consultant and a bunch of strangers on the Internet?

The images are extraordinarily sad to witness. The texture of late-1990s digital photography, with its poor dynamic range, burned-out highlights, and pixelated speckling, produces an effect akin to that described by Lucas Hilderbrand in his study of analog videotape as the "aesthetics of access." It's the texture of bootleg tapes accumulating the traces of their reproduction, the poor quality that implies the intimacy of amateur labor.[23] In both form and content, these are in the space of private life on the one hand and private fantasy on the other, with the boudoir's stage properties of drapery and Ionic columns. One instantly regrets having seen them and sympathizes with the people thus exposed. To their credit, the antispam activists of news.admin.net-abuse.email, who were alerted to this early on, expressed deep unease about both the methods and the choices of the MITW—while painstakingly dissecting the data that turned up and mirroring the site to keep it online.

The data made available to them, setting aside the corner of this project devoted to public humiliation, was unprecedented: a huge mass of documentation for the quotidian work of a spam operation. The chat logs, in particular, were a goldmine—they read, from an antispam perspective, like the transcripts of a gangland wiretap. There is overt discussion of finding disposable Internet access accounts with service providers (to skip out on once the complaints start coming in), spoofing return addresses, lining up client work, and the logistics of keeping the spam flowing. We will never get anything closer to an ethnographic account of spamming prior to 2003—before the CAN-SPAM legislation, before filtering and botnets and the war of the algorithms—than we do in these files (fragmentary as they are, regularly referring to exchanges to which we do not have access, conducted over email or the phone). These 5 MB are like a geologist's road cut, a sectional slice across the enterprise of a midlevel spamming crew hard at work.

It is a profitable business with repeat customers. There is much back-and-forth about people sending checks by Federal Express (FedEx), and negotiations about rates: "How much for 400K a day for the next 4 weeks?" asks a potential client. "300k max a day—24 hours of mailing—dedicated machine. $500.00/day," Garst responds. When the client demurs at the cost, Garst draws on her own experience: "Guess it depends on how much can be made from that mailing. 300k nets me much more than $500.00 when I mail for myself. I'm assuming a client would make much more than that if he or she has the right product."[24] She and her employees gather mortgage leads worth $8 apiece, netted with state-by-state mailing campaigns (two of her employees toss titles back and forth: "What would you use for a subject . . . just something like "Attention Homeowners!!"?" "I've used things like / Pay Off Everything / Pay It All Off").[25] Deals are assessed informally. "I now have that mortgage deal, cable boxes, anabolic steroids and Adult . . . cable box and steroids pay 20 percent of sale. Those are proven and have been selling on net for a long time. The mortgage I can get 2.50 per lead" offers an associate of Garst, based in Las Vegas.[26] The mailings are run off of Garst's suite of machines or subcontracted out to remote employees who are sent files, ads, and instructions. These employees, like Garst herself, lead a strange version of the freelancer's life: "Then [I'm] going to the bank where they think I'm doing illegal stuff 'cas [sic] I'll come in to deposit checks from $500–$100,000 dollars, plus $300 worth of dollar bills."[27] These discussions in the chat logs are interspersed with a great deal of personal conversation: vacations, boat purchases, various romances, the blow-by-blow of buying a house and attending a high school reunion. Sorting through the logs, and a number of the other files, produces a strange and perhaps unavoidable narrative empathy for the people whose conversations we read. Will Valentine finally get a place of her own and no longer have to live with her boss, Garst? In the midst of an address-gathering project, grouping the addresses by state to sell the leads to mortgage brokers with appropriate licenses, Valentine writes: "Gotta go help search for Rodona's dog. Be back asap."[28]

Under the permanent project of getting "good numbers," the forces that govern the life of the spammer can be discerned and separated out by going through all the files carefully.

ACCESS TO RESOURCES

This issue is the constant constraint, the crimp in the spam production process. They need bandwidth, clean addresses to send spam from, relay servers to disguise the source of the messages, and website hosts for the pages where people can fill out order forms. These crucial resources are constantly being lost and regained. One aspect of the explosion in commercial ISPs is a proliferation of free trials, which spammers can sign up for, use to move a few million messages, then drop before the complaints come in and the account is shut down. Garst's crew grab slow dialup accounts when they must, but they much prefer the use of Garst's high-speed line, when that isn't being blocked for misuse. Everything is done on a temporary basis, one step ahead of the complaints, flames, and inevitable plug-pulling. "4 mil a day per box, want 25 users, 100 mil a day, give them a few days and it'll all be down." Try European dialup service providers, suggests Garst's colleague, and pay the five cents a minute for the call. "I've tried the small ones as well. They work one time and then they get smart and that's the end of it."[29] There are mysterious lags and slowdowns; outgoing messages slow to a crawl, or the response rate abruptly drops to levels far below where it should be—the mail is still going out, but it is being stopped somewhere on the way.

Everyone's dream is an understanding (that is, corrupt) service provider who will look the other way and ignore or somehow sidestep complaints. The spammers call this arrangement "bulletproof hosting"; the antispam community calls such a deal a "pink contract." Garst, increasingly desperate during a difficult period, looks into cutting a deal with a Chicago ISP for $5,000: "Backbone has no problem with the content . . . even adult. All of the folks there mail only AOL so the heat will be less than if we were mailing GI [the "general Internet," email addresses across all domains rather than only AOL addresses]. It's worth the gamble to me." She had received a full-dress tour there: "We talked to everyone from the General Manager to the NOC [network operations center] manager to their legal department."[30] The need is for some form of access at the infrastructural level, a space on the network with a role equivalent to that of failed states for drug smuggling, with a reliably open airstrip and a flag of convenience.

Then there is the issue of getting good addresses to send from, especially within AOL's network. AOL accounts were valuable, with an address from which you could send messages to the rest of AOL's network. They provided ready access to an enormous population of users more likely to be sheltered and unfamiliar with the network's threats, easier to exploit, and less likely to make trouble outside the precincts of AOL. (Addresses to send *to* were also, of course, valuable. McWilliams notes the price Davis Hawke and his protégé paid for the full database of AOL user accounts, bought from a guy named Sean who said he'd bought it from a disgruntled AOL software engineer: $52,000 for 37 million customer accounts, each with multiple screen names. The addresses were that valuable both in themselves and for eventual resale to other spammers.)[31] But of course AOL would terminate your account for spamming other users once the volume of complaints got big enough. Hence the practice of "fishing," as Garst's employees refer to it—with usable accounts noted in chat with fish symbols, "<><"—or, as we know it now, "phishing," misleading people into giving you their passwords. (The term's first known appearance is in the Usenet newsgroup for the hacker magazine *2600* early in 1996, but it is referred to there as a preexisting slang term; in that context, it also refers to collecting AOL user accounts.)[32] They worked in the phishing mode we recognize today, with an email purporting to come from AOL requesting that the customer to update his or her login information in return for a free month of use. The MITW's collection includes the landing pages they set up for the login credentials, which look depressingly similar to the pages linked from current phishing messages seeking bank and PayPal credentials. The login information collected this way provides one-shot accounts, good for a single burst of spam before they get shut down, and the supply is constantly running low. The lament of the early morning spam campaign: "freestation won't let us change/create, crosswinds shuts 'em down so fast, I can't get the right connection w/ a★★net, and I don't want to run out of fish."[33]

TACIT OPERATING KNOWLEDGE

Much of the work of spamming is built on informal knowledge exchange. Obviously, there is no process of formal education or certification, no conferences or refereed journals—those PDFs with titles like "Make a Fortune Marketing on the Internet!" are about as good as it gets. No one is doing much math or software development beyond keeping an eye on the ratios of how many messages go out and how many replies, new leads, and "undeliverable" messages come back. Instead of formal analysis, there are rules of thumb ("[the weekend is] when smut does well") and a constant exchange of low-level tips and instructions. They share notes on producing HTML pages, automating processes that make messages harder to trace, issuing commands to the server, dealing with baffling and buggy spamming software, and, as mentioned previously, finding and keeping good connectivity.[34] Computers are largely mysterious and frustratingly recalcitrant appliances, "and our tech is flaky as hell."[35] The conversation is rife with guesswork and reverse-engineering discussions trying to figure out why something is happening. There is a thrifty inventiveness apparent as they string together networks of free and stolen services to produce a constantly changing spam apparatus that can discard some functions and substitute others as they get banned or busted. In this regard, the spammers offer a strange symmetry to the volunteer antispammers in organizations such as NANAE, many of whom, incensed by the flood of spam but unfamiliar with the technology, engage in similar kinds of skill-sharing, hints, and how-tos.

OTHER SPAMMERS AND CLIENTS

By and large, this tacit knowledge-sharing works in a collegial environment—not just within the company, but between companies and groups of spammers. They broker deals for addresses ("new lists I've built are att.net, gte.net, usa.net, hotmail.com, etc. . . . a bit over 1 million that have been built recently [say last two weeks]"), gossip about others in the field, and make recommendations about spamming software and how to get it at discounted rates or for free.[36] There is a weariness with the venality, distrust, and universal loathing entailed by the business. After establishing trust through "Doc," a mutual friend, Garst and another spammer dubbed "GateKeeper" open up about their experience:

```
GateKeeper 1/18/00 6:01 AM ;o) been in the business long

Rodona 1/18/00 6:02 AM 4 1/2 years now. You?

GateKeeper 1/18/00 6:03 AM just about a yr longer . . .
feals [sic] like 15

Rodona 1/18/00 6:03 AM Isn't that the truth:-)! Getting
harder and harder to survive. . . . I hate that it has come
to this—where we can't share what we learn, etc. with one
another like the old days.[37]
```

A few weeks later, after they do a handshake deal for an address verification system, Garst writes, "I used to be so disillusioned by others in this business but what goes around comes around and I don't want the bad stuff coming around to me so I try to keep only the positive stuff flowing:-)." Despite being painfully anodyne, this sentiment captures the strange milieu of spam at this moment: people in a complex relationship with legitimacy, accepting the cost of getting ripped off with some regularity in return for being able to operate largely outside the space of courts and contracts, in an informal para-criminal economy.[38]

Spammers exploiting one another was, and is, a common feature of the trade. Garst and yet another fellow spammer chatted at length one night about a third with rumored substance abuse problems, who seemed to be making big, dubious promises about connectivity and servers for bulk mailing and looking for investors.[39] The same is true of clients shirking their payment for a campaign. Her problem with a client whose check bounced and who then accused her of stealing his advertising text was carefully thought through: "How can I effectively shut him down? I'm tired of messing with him," Garst asked, to which her fellow spammer answered (after asking about the kind of shutdown she meant: "FCC, FBI, Hacker . . . ?") by describing the rules of their informal system.[40] The client, one of the mortgage brokers, could threaten to report Garst for her actions, but she could in turn report him, likely costing him his license. She could threaten to launch a spam campaign to cast a bad light on his business, for which he would have no reprisal. (Notorious spam campaigns had already been launched to suggest that a deadbeat customer or overly vocal complainer was, say, a child pornographer or something equally despicable.) Everyone involved in her side of the business was just shady enough that they usually maintained an equilibrium of mutually assured destruction.

ANTISPAMMERS

The truly unruly force were the antispammers who attacked her work at every turn. The language Garst and her employees use to describe the "antis" is full of irritation and contempt. "Looks like I had a zealous little anti last night . . . poor loser," one wrote in 2000. This anecdote echoes the recurring language used by spammers to paint their opponents as geeky fanatics, from one of The Wizard's employees in 1998 ("Boy, some people got a lot of time on their hands . . . Tell [the antispam protestors] to get a life") back to Canter and Siegel in 1995 ("the wild-eyed zealots who view the Internet as their home. . . . They have a very exaggerated sense of importance of the Usenet in their lives, that the average person doesn't share").[41] We are normal people, say the spammers, trying to make a living, not like these "anti-commerce net-nazis." We are being harassed by obsessive nerds who are "anal," "extremists," "flamers," "evil," "bad guy"—this latter language coming from the running bulletins of one of Garst's providers of fresh addresses, who also passed along new names for their list of antispammers.[42] The function of the list, which grew to enormous size—more than 200,000 addresses—was to exclude those addresses from spam mailings because they belonged to troublemakers.[43] The spammers set them aside to avoid reprisals. Some people got special mention, such as "ANTI EXTEMIST & FAX BOMBER / ron2ron@earthlink.net / fax terrorist."[44]

The antis were deeply annoying but did not actually function as a deterrent. At one point prior to the exploit by the MITW, "We got hacked . . . got posted on an antispammer site today with ALL of Rodona's info—phone number, name, iCQ #, PO box, an old AOL address."[45] She just updated her information. "Please don't pass that along," she wrote after sending her new number to her address provider. "I'm trying to start fresh:-)!"[46] Getting bogus magazine subscriptions and angry calls didn't even slow the spammers down, though they kept a close eye on their charivari: "can you send me how to get to the NANAE newsgroup? THANKS!!"[47]

Most of the legal action undertaken by the ISPs was similarly an annoyance that didn't deter business. AOL's lawyers periodically sent out letters in a strange demonstration of diligence, requesting information about how mail is sent and with what programs but never following up. Garst handled it with sangfroid: "Just respond that you bought program anticipating doing some marketing for your online business but decided it was way to [sic] much trouble. That's what another girl did and she's never heard a word from 'em since."[48] (Though it was possible to overdo it with AOL—the spammers weighed the possibility of a "raid . . . with warrants" if enough messages were put into the AOL system.)[49] She is astonishingly blasé about the possible legal consequences of her stock-pumping schemes: "Since I have an inside of sorts:-), it seems it would be wise if I purchased some stock that we are promoting."[50]

There was one legal challenge that was genuinely problematic for the spammer community, which was mounted by Paul Vixie. Vixie, a DEC employee, major contributor to the Unix operating system and the underlying Internet protocol, and author of a number of RFCs, launched an antispam nonprofit in 1998. It was built around a complex system of blacklists—the Mail Abuse Prevention System (MAPS). Where the threat from Vixie was felt most immediately was not in his nonprofit, however, but in his capacity to strike at the spammers' most vital resource by working with service providers to keep them from getting any significant access to bandwidth. The central point of failure in the plan to set up servers with a spam-friendly service provider in Chicago is "that much mail into gen internet ["general Internet" as described previously: addresses outside AOL] is gonna attract vixie big time and it will be over, doubt they'll get 30 days . . . Vixie is gonna see thru this and deal with the backbone [the service provider] harshly. Same old story vixie is quick death"[51]. What was truly problematic was not the swamped and slow Federal Trade Commission (FTC), which was receiving many thousands of complaints every day in its "unsolicited commercial e-mail" inbox (more than eight million over four years) or the many outraged individual antis whose hectoring abuse was simply part of doing business. It was strategies that kept the ISPs from giving the spammers connectivity.[52] It was an attack on the critical infrastructure.

Though the law moves slowly, it did eventually catch up with Garst and her colleagues because of the pump-and-dump stock activity, bringing proceedings against both her and her client under the Securities Act and the Exchange Act in 2002.[53] (The order keeps the odd network slang in quotation marks: "Between approximately September 1999 and May 2000, Garst disseminated large numbers of unsolicited 'spam' e-mail messages touting four stocks.") Many of the stories of the most prominent spammers, from Garst and Sanford Wallace to Oleg Nikolaenko—who was arrested in Las Vegas in 2010 while visiting an auto show—end with the law, with settlements, civil judgments, and occasionally prison time. There was Wallace's hundreds of millions awarded to Facebook and MySpace; Hawke's almost $13 million owed to AOL; Christopher Smith, $5.5 million awarded to AOL and thirty years in prison for charges related to drug dealing, fraud, and money laundering; Leo Kuvayev fleeing North America (owing a multimillion fine in Massachusetts, hurriedly shutting down an office in Montreal) and living in hiding. What is striking, in retrospect, about this glimpse into the operation of spam is how simultaneously secure and precarious it is. The spammers are operating in a space of astonishing abundance, one in which fresh resources constantly appear to replace those lost, where new workarounds can be crudely stitched together and then shared via chat. It's just a banal job, without any real fear from the "antis" or the police. The people that worry the spammers are the other spammers and their clients. Yet this job is always changing, being rebuilt as it goes on from day to day.

From the perspective of the diminution of spam, it's notable how little effect the volunteer antispammers had, given the extraordinary volume of their outrage and energy of their efforts. They expose the very real limits of "shaming and flaming" when directed toward the shameless and secretive. Setting aside the problem of actually stopping spam, however, the process of antispam activism remains a decade-long case study in online collaboration, community work, and negotiation at the barricades. In the next section, we focus on antispam, starting with the first wave of technological methods (the cancelbot wars and their strange outcome), the development of a legal toolkit for thinking about spam, and finally the creation of NANAE: the charivari in power.

BUILDING ANTISPAM

THE CANCELBOT WARS

If the capacity of networked computers for multiplying and distributing information enables the production of spam, can it also solve it? There's a pleasing symmetry to this arrangement, a *pharmakon* in which the poison, properly applied, becomes its own antidote. As Usenet struggled with the spam boom in the 1990s, a method to eliminate messages already existed, having been built to address the problem of misspeaking and regret. It was called the "cancel" message. If you posted a message and you wanted to take it back, whether an accidental duplicate or an irate flame, you could send another message to a special newsgroup called "control" requesting that your message be removed from the feed and not passed on. The cancel message simply needs to come, or appear to come, from the message's original author and use a given message's unique ID number to begin the cancellation process. As the cancel message moves around from machine to machine, it can take care of removing your message, whether it arrives before or after the message you want removed. A similar system had been developed for The WELL, allowing users to "scribble" their posts, removing them from the board (while leaving the trace that they had spoken)—with the downside that a user in a huff or on a self-destructive binge could remove his or her own prior speech from existing conversations, leaving the other participants arguing with a phantom and a conversational record of only one side of a discussion, like a telephone call overheard. Canceling on Usenet had similar problems but developed a whole new set of issues when it was turned into an enforcement tool to stop spammers.

The problems began with authorship. The whole point of a cancel message was that the individual author could manage his or her own statement. To deal with spammers, we have to wade into the territory of the "forged cancel" or, more euphemistically, "third-party cancel" and use a method that allows us to pretend to be the original source, choosing to revoke our words. (This trick is done by duplicating the original author's "Sender": heading.) A set of structured criteria was created to manage the third-party cancellation process in order to keep it from creating too many social problems, starting with the header of the cancellation message. The "Sender": is a duplicate of the original author's, but the "From": heading is set to that of the person canceling, who is therefore available to take complaints and flames if they have canceled someone else's message inappropriately. The same applies to a special header, "X-Cancelled-By":. The body of the message has no technical bearing on how the cancellation process works, so it contains an explanation of why the third-party cancel was issued. To quote Tim Skirvin's example (we will encounter Skirvin again soon, at the founding of NANAE):

```
WOODSIDE spam cancelled by clewis@ferret.ocunix.on.ca

Original Subject: Sell YourPhotosNYC.Agency

Total spams this type to date: 1.758

Total this spam type for this user: 1041

Total this spam type for this user today: 503
```

```
Originating site: Internetmci.com

Complaint addresses: spamcomplaints@mci.net
postmaster@mci.net[54]
```

Accountability, therefore, is built into this process at every line: here is who canceled the message, and here is why, and here is the case against the original sender. It has a strong resemblance to the explanatory structure of Wikipedia's editing process. Bear in mind, however, the difficulties Wikipedia has faced over the cultural struggles between "inclusionists" and "deletionists," in which the merely expository system for describing why an article has been deleted soon became a flashpoint for larger negotiations about what Wikipedia should be—a conventionally respectable *Britannica* on a screen, or an omnium gatherum reflective of all information, rather than only that which survives the editing process for scarce paper inches?[55] Parallel problems soon appeared in the Usenet message cancel process.

"It seems to me that the real issue at stake is what the purpose of the Listserve [sic] groups is," writes Jason Webb of a notorious case of message cancellation involving Libby Hubbard, whose online presence is largely under the handle Doctress Neutopia.[56] (A listserv is an automated email list with an archive that is used to share information and conduct discussions.) Hubbard became mildly infamous for producing something akin to the description of Minnie's "spam" in LambdaMOO—legitimate contributions that were couched in a style some others found confusing, confrontational, irritating, or counterproductive. (It largely took the form of proselytizing for a philosophical and political agenda in an idiomatic style with many neologisms, like a mixture of Mary Daly and Buckminster Fuller.) This approach led to "many occasions" when Hubbard was "'flamed' for my alternative Neutopian Vision, not by one individual, but by various Usenet and Listserve [sic] groups." Many of the other contributors adopted versions of the killfile technology described previously to stop receiving her contributions. It also led to a far more serious move, though, on the part of a listserv administrator who deleted all of her posts retroactively. Tamir Maltz's study of customary law online describes this event as "an Orwellian erasure of history" and captures the strange resemblance of such an act to the redaction of murdered or disappeared Soviet political figures from photographs and records (on, obviously, a different moral-historical scale).[57] Though it is rather too grandiose to compare a classic online flare-up—an eccentric contributor and an increasingly impatient community at odds—with an icon of modern barbarism like the "memory hole" in which the censors dump documents and records in *1984*, the capacity to simply cancel or erase an author's contributions was deeply disturbing. It touched on the fundamental instability of digital information, the strange fragility by which any trace that might mark that something had been here is gone. Hubbard/Neutopia would not be left even the chisel marks made by the erasure of the Pharaoh Hatshepsut from the records of walls and monuments. Such use of cancellation and deletion tools created the fear that this online environment could be Orwell's incinerator, which didn't leave ash to suggest an omission in the official record. (However, copies might live on in the archives of individuals who had been on the list—or, in the case of Usenet, in a subtler set of archives of the deleted, as will be described.) Capacity to cancel and erase someone else's messages was clearly a social problem to manage, not a straightforward solution in itself.

Though the system of cancels was full of painstaking justifications and methods for accountability, a deep concern about "mission creep" and misuse weighed on the cancellation project. A number of prominent newsreaders and hubs for distributing Usenet messages either accepted only certain forms of cancel messages or refused to accept them at all, whether from a simple anxiety about misuse or as an expression of a priori ethical commitments to free speech. The exemplary "site of virtue" run by Dave Hayes not only did not accept third-party cancel messages but didn't pass them on so that newsreaders dependent on his site could get a less-filtered version of the network.[58] These concerns went double for automated cancellations or "cancelbots," which were set up after increases in the flow of spammy or simply overposted messages to automatically send out cancel messages to eliminate them. Cancelbots managed to combine anxieties about coercion and moderation-through-erasure with the potential disaster of automation, a system that lacked a human capacity for contextual awareness and nuance and which could—and did—go haywire. Recall anon.penet.fi, the anonymous remailer for Usenet described earlier, which broke the flaming-and-shaming system of deterrence. It inspired Dick Depew to build the Automated Retroactive Minimal Moderation program (ARMM), which would automatically issue cancels for messages from anon.penet.fi, and produce a follow-up message—and follow-ups to the follow-ups, recursively: hence its first time out of the gate leading to the charge that ARMM itself had "proceeded to spam" Usenet. Although it was simply poorly coded, it spoke to broader concerns about unleashing automatic erasure systems on what was supposed to be a thriving environment of open conversation.

Then there was the problem of who got to be in charge of cancels beyond the most obvious cases of spam: what was the recourse if bullies, trolls, or other bad actors got ahold of the forged cancel technique? As Gabriella Coleman has documented, the rough-and-tumble ethos of free speech and public discourse at all costs that formed a key element of both hacker culture and the network public sphere had a near-perfect antithesis in the secrecy-obsessed lawsuit-and-intimidation culture of the Church of Scientology.[59] This crisis continues into the present day, with cease-and-desist letters in response to documents posted online and the mass organization of Anonymous's street protests against Scientology in the winter and early spring of 2008—but it began on the Usenet newsgroup alt.religion.scientology in 1994. Already a scene of confrontation and much vituperation between activists and staff of the church, the battle in the group escalated when messages began to vanish late in the year. No one was explaining why the cancels were issued, and a graduate student in biology named Chris Schafmeister built an ingenious tool that compared the logs of messages received at one of the sites that refused cancel messages, and the log of cancels, with their unique ID numbers, to create a record—and a public alert—whenever a message to alt.religion.scientology was canceled. As Wendy Grossman describes, these doings set off a long cat-and-mouse game between the censors and the activists.[60] It culminated in the operator of that same anon.penet.fi, which had been used to anonymously post information about Scientology to the newsgroup, being raided by Interpol and giving up the identity of one of the users of the remailer. The capacity to cancel messages, even for a seemingly obvious case like spam, was not the unambiguous good it may appear to be at first. What qualifies as spam and who decides remained very much at issue.

Getting back to the erasure of Hubbard/Neutopia's messages, the rejoinder to her essay from Jason Webb includes that uncertainty as to what a listserv is for: "if the purpose of the group is to promote the free expression of ideas they are not succeeding. . . . We have to be active in trying to create an environment where all ideas can be expressed without the fear of being ostracized." But is the point of listservs, Usenet groups, email, and the Internet as a whole to "promote the free expression of ideas"? That certainly wasn't the purpose of its initial construction, and those goals remain a deeply complex and problematic instance of the language of ideology in the debate over what the Internet is to be and which constituency gets to decide—one of the root paradigms at work. One of the manifestations of this uncertainty is the struggle over metaphors, a struggle with very practical, physical, and technological consequences. What was spam *like*? To decide the answer to that question was to decide indirectly what the network itself was like and how people and their machines should behave on it. This question was played out to great effect in the shaping of law.

SPAM AND ITS METAPHORS

Junk mail, unsolicited telephone calls, postage-due mailings, direct marketing, people crashing parties and shouting, trespassing, underregulated businesses, free public speech—and, more abstractly, parasites in ecosystems, abusers of the commons, noise in the public sphere, trash: these are some of the metaphors with which various groups tried to get to grips with spam.[61] Metaphors are powerful, carrying a freight of argument, controlling the territory in which negotiations are made ("Do you know what 'metaphor' means?" asks one of Latour's alter egos. "Transportation. Moving. The word *metaphoros*, my friend, is written on all the moving vans in Greece").[62] One of the remarkable things about the project of building antispam was how hard it was to build workable metaphors that could explain to lawmakers, outside authorities, and the general public what spamming meant and why the spammers were a negative influence.

"Imagine someone who's checked into a hotel that offers free local calls making 6,000 calls to you on your cellular phone. They don't pay a dime, and you have to pay for listening to advertisements you don't want": so goes an amateur press release drafted on Usenet after Canter and Siegel, by a writer with a .edu address, reflecting membership in the old dispensation.[63] Quite distinct forms of communication are drafted into this bizarrely improbable scenario, which fails as a metaphor in that you're under no obligation to take calls, much less listen to them, coming from that number. The argument that is struggling to be made there is clear—there's an enormous difference between what spammers pay to send and what everyone else pays to receive—but the problem of explaining that to a mid-1990s general public leads to bizarre comparisons. "USENet is like a convention center with thousands of meetings going on all at once. Anyone's free to drop in on whatever discussion interests them. It's like C&S hired an army of people to stick their heads in every doorway and all scream 'GREEN CARD LOTTERY' all at once." Is it really? The implications of each are clear: the owner of cell phone can claim the hotel caller to be a nuisance, and meeting rooms for a conference imply relative privacy, discourse, and certain strictures of humane conversation. The press release comes back to that one: "To return to the convention center analogy, they've warned us that they're going to hire yet another army (and tell other people where to find one) to storm the group meetings. So we're trying to find ways to bar the doors." Is it like a phone call, or like a group of people in a physical space? But with neither can you filter statements the way you can on Usenet. With neither do statements propagate outward, copy by copy, at no cost to the sender and a very small cost to everyone in the chain that aggregates into a massive expense overall. Consider, as well, where the analogies lead: if the plan is being explained to the notional public as "barring the doors" to the conference rooms, the antispam project begins to sound much more like the work of the "zealots" Martha Siegel mocked, who want to seal off their little fort from the public and control the communication that can happen there.

The temptation is very strong to think of Usenet—and the Internet more generally—in terms of *place*, and spamming as a violation of the boundaries of place. This impulse can be seen in the discussion around privacy online, for instance. A substantial literature is devoted to analyzing the Fourth Amendment of the U.S. Constitution, attempting to either draw on or disentangle it from the language of home, threshold, street, and wall—those places where you might have "a reasonable expectation of privacy" and where you assume the possibility of being overheard. When your voice moves beyond the precincts of your home, over the telephone wires, do you still expect privacy? When you are posting messages from a terminal that will circulate around the world, under what circumstances can you constrain other writers who may be in other states or countries, writing under different assumptions of use? In a final, deeply and tellingly strange comparison, the press release draft quoted previously ends with a comparison that draws on layers of mores, written law, architectural space and design, and social understanding: "People are not trying to harm CSLaw or Canter and Siegel. All they're doing is asking that they not be allowed to violate the USENet community. If it's vigilantism to ask that people confine their activities to designated areas, then every business and school and hospital in the country is run by vigilantes, too." Usenet is not a place, however, and certainly not a place like a business, a hospital, or a school, whose designated areas and confinements develop out of powerful formal systems and constraints at large in human societies. A hospital is the immediately visible and tangible expression of licensing and funding projects, sterility and fire safety requirements, nursing unions and patient's rights organizations, medical schools and residency programs, endowments and philanthropy, equipment procurement relationships, schedules and hiring practice, diagnostic consensus as to inpatient and outpatient procedures—to say nothing of insurance and health care mandates, which are national and supranational architectures whose influence manifests at every level of a hospital's day-to-day operation. Usenet, by contrast, is a protocol for handling the storage, retrieval, and forwarding of messages among a set of computers in a network. As Latour points out, the question of where you draw the line around something—where are the edges of IBM, for example, or a city's police force, or for that matter a telephone?—is already a critical analytical decision, despite often taking place prior to the analysis as such.[64] Where you draw the line has a lot to say about the kind of shape you expect. A hospital is a building, but it's much more than that; so is Usenet a protocol, but it's also much more than that, as the press release struggles to express and as Canter and Siegel, in their crudely opportunistic way, clearly discerned. It's a protocol for a set of computers, but it's also twenty million people, from all over the world, in six thousand active discussions and millions of words, without any formal law or proprietary control.

The metaphors that are ultimately settled on can have powerful consequences. (The third section of this book, which is full of worms, zombies, bots, and metaphorical mushroom clouds, describes in detail some of these eventualities.) "All technologies involve 'scripts,'" as Jessica Johnston reminds us in her ethnography of the computer antivirus industry. She illuminates how the scripting of different technologies—embedding them in larger cultural and political agendas and narratives—is rendered much more visible by the moments when their metaphors could have been different.[65] Consider the computer "virus." The virus is associated, obviously, with existential threats, both in the mass graves of influenza and future avian flu outbreaks, and with the ontologically uncanny para-life of the virus itself, a complex molecule neither quite matter nor living thing. It ties into metaphors of health and illness, infection and pandemic, the terror of death and a deeply unsettling management of nature (think of white-HAZMAT-suited CDC professionals torching huge piles of dead chickens with flamethrowers) by an elite corps of professionals. All of this fear is brought into the script implied by the language of the computer virus, but a "virus" was not the only possible metaphor for the technology. "Alan Solomon . . . a veteran antivirus researcher with a PhD in economics, critiqued the virus metaphor, suggesting that this medical/biological metaphor of 'virus' is 'too emotive' . . . Instead, he proposed 'weeds' as a more appropriate concept for describing the threat of computer code."[66] With "weeds" comes a very different culture of metaphors, of strong and weak ecosystems, each person cultivating their own garden every day to keep invasive species at bay. It is a much better metaphor for expressing one of the global computer network's key points of weakness to "viral infection": the monoculture of computers running the Microsoft Windows operating system, often poorly patched and unmaintained by users, making the network as vulnerable as the cloned Cavendish banana trees are to fungus attacks. Without overstating the influence of metaphor, it's striking to consider how much that nomenclature might have changed the practices of security and programming around self-replicating computer code: computers as gardens rather than bodies, with diverse software populations to be tended and pruned by attentive and self-reliant users, potentially capable of weed resistance in their interdependence, with the professionals as agronomists, breeders, and exterminators rather than doctors at the cordon sanitaire.[67]

The choices involved in building metaphors about spam had long shadows and complex implications. Are spammers free riders, and the network's story the tragedy of the commons? Are they naturally occurring parasites in a complex ecosystem? Do they engage in attention theft, or are they a tax on the stupid and gullible who can't take care of themselves? Daring entrepreneurs in an underregulated environment (that is, a space into which the government must move)? The inevitable collapse of a free system into abuse and necessary governance? The consequences of these metaphors run from the particulars of how you would build an antispam system to how much of a price you're willing to pay for it. Is spam a headache you can live with in the name of free speech, or an invasion that must be halted and exterminated, even at the cost of dramatically changing the network's existing culture?

An immediate metaphorical precedent suggested itself in law: the 1991 Telephone Consumer Protection Act (TCPA), developed to manage telephone solicitations, especially the provisions about "junk faxes," unsolicited advertising sent to fax machines. Many spammers, including Sanford Wallace, also dipped into the junk fax business. "They are doing a press release and FAX BLASTING!," writes Garst of one of the pump-and-dump stock schemes.[68] "As well as 10 mil emails?" her collaborator asks: the two forms of advertising were often closely allied. (The infamous company fax.com was moving almost 800,000 unsolicited faxes a week by 2002—approaching spam numbers for a paper operation.)[69] Drawing on the TCPA had a certain admirable simplicity that was also its weakness: it made the businesses sending the spam responsible and liable. That means you have to track the business down, get the perpetrators to come to court, and find their funds, which is far from impossible but doesn't scale to the rapidly growing population of spammers. Tracking down Garst might be relatively easy, but is it worth the FTC's trouble to send off a search through the backwoods of Vermont for a potentially armed neo-Nazi whose profits are buried in caches of precious metals somewhere? Multiply that extreme case into hundreds of situations of varying difficulty, obscurity, and intricacy, with more new adopters of spamming technology and techniques coming online every day. ("Spamming" had not yet been clearly defined in any case.) Many spammers ended up being based in Florida because, along with sun and scenic beaches, it offered greater protection for real estate from civil judgments and a number of gentler antifraud statutes than other states. There were thorny jurisdictional issues as well. Many spammers were transacting herbal medications and other health-related products across state lines, engaging in stock touting and insider trading, selling pornographic materials, laundering and hiding their difficult-to-document profits, running dubious or entirely fraudulent philanthropic organizations—or all of the above! So, to speak only of U.S. agencies, is the FCC, FDA, FTC, SEC, or FBI obligated to deal with them? And on whose behalf? Users of the Internet could often be a self-marginalizing constituency with the most vocally political also the most uncertain about what role, if any, they wanted the territorial government to have. Besides, spammers themselves were quite capable of arguing that they were legitimate businesspeople being harassed by childish pranksters; they launched suits and legal campaigns of their own. Paul Vixie's MAPS project, the one Garst and her address supplier were worrying over, were so used to this argument and ready for a decision in their favor to set an antispam precedent that they had a page on their site with instructions for "How to sue MAPS."[70]

More abstract, and graver, were the problems of speech and freedom. Spam, despite being rather less compelling a case than the publication of *Ulysses* or the declassification of the Pentagon Papers, touches on issues of speech everywhere. The problem with a number of different bills proposed to fight spam was that they built their metaphors on more tangible media such as faxes and postal mailings, which, when translated in the electronic sphere, had unsettling and potentially unconstitutional implications. HR 1748, the "Netizens Protection Act of 1997," was directed specifically against "the unseemly practices of the junk e-mailer"[71] and would open spammers to suits of $500 in small claims court for unsolicited messages, or $1,500 if the spammer could be shown to have knowingly violated the law. (Spammers would be obliged to include their contact information in the message.) This was an amendment of the part of the Telephone Consumer Protection Act of 1991 devoted to junk faxes—using it against spam was "a natural extension of existing law," which seems logical enough. This case is another, surely, in which the recipients are obliged to pay for an unsolicited message. There is existing language about the "captive audience" who cannot refuse the ad, with an opt-in provision. Many antispam activists found various problems with the bill, but Brad Templeton summarized them most deftly.[72] Such a law could change the culture of email for the worse. It attacks not the *bulk* of bulk mail, the massive volume, but is instead based on the content of individual messages. The criteria are wrong, and wrong in a way that could have a chilling effect on the medium as a whole. Is an automatic signature at the end of a message (with, say, contact information and a slogan for one's business) potentially classifiable as spam, should someone want to make trouble for you? There are plenty of scenarios as part of regular experience in which an unsolicited message with commercial content, from a friend or a stranger, is actually welcome; such communications are obviously different from spam, but it is unclear if they would be so in the eyes of the law. And, as Templeton points out, will you have to fly to Anchorage to defend yourself in an Alaskan small-claims court if there's a misunderstanding about something you sent—or you are the victim of simple malice?

The story of AT&T's suppression of its own invention of magnetic tape audio recording offers a good parallel for the problem of the spam-as-fax metaphor and the consequences of legal uncertainty. In his study of the monopolistic dynamics in information technology, Tim Wu draws on historian Mark Clark's research to document a remarkable object: a cabinet-sized device built to use magnetic tape to allow a telephone caller to record a message. It is an answering machine built in the 1930s under the auspices of AT&T. Engineers and inventors at the company did an astonishing job with developing audio recording technology on magnetic tape, which the company then suppressed for decades.[73] Wu's point with this history is to outline the cycles in which innovative technology companies come to slow and even halt innovations, but the particular reasons for AT&T's withholding of the technology capture the trouble of crudely managing spam with the law. AT&T's executives feared that if the technology of magnetic tape recording became well known and widely distributed, use of telephony would sharply decrease because of the suspicion of surreptitious recording. Businessmen would be afraid of being taken to court for promises or claims made over the telephone, and time on the line spent discussing things personal or even obscene (a not-insignificant use of the network) would be curtailed. Bills such as HR 1748 applied to email would have a similar curtailing effect to that imagined by AT&T, but as a practical matter rather than a paranoid fantasy. It would be extraordinarily easy to litigate over a message, and email would become a format in which people were conscious of being potentially in public—indeed, in court. Furthermore, as Templeton points out, there are other possible solutions, many of them technological, that haven't yet been deeply tried and that do not offer such an immediate prospect of turning email into a minefield. (That bill, and a number of others like it, died in committee.)

If you can't sue the spammers for all the reasons of onerous complexity and jurisdiction outlined thus far, who else can you attack? Where are the points of failure in the spam machine? As described, spammers need compliant, or at least inattentive, ISPs, and ISPs are stable, overt, and eminently suable. You can make the ISPs liable for the spammers on them, though this approach of course opens up another legal nightmare of communications providers rendered liable for the actions of those using their facilities—AT&T does not face penalties because mobsters make use of telephones. It could be potentially ruinous for ISPs: S 771, the Unsolicited Commercial Electronic Email Act of 1997, required ISPs to block commercial email within 48 hours of a customer complaint or face thousands of dollars in fines. Spammers, who are not exactly scrupulous when it comes to using stolen accounts or open relays to send messages, could thus potentially bankrupt an unlucky service provider. As an alternative, you could enable the ISPs to sue the spammers, based on an interesting and unexpected analogy for spam. Everyone is conscious of spam as being akin to telephone solicitations and unwanted fliers (recall Postel's use of "junk mail" for a slightly different case), but what about spamming as a form of trespass? "An ISP's network is like its real property," writes Timothy Casey in a guide to liability for ISPs (a phrase that will soon seem antiquated in its distinction between "real property" and "network").[74] "Want to know where the action in a culture is?" asked Stewart Brand in the late 1980s. "Watch where new language is turning up and where the lawyers collect, usually in that sequence."[75] In books such as Casey's, practical and hands-on manuals for people running an ISP, we can see both the language and the lawyers in motion—a network on its way to becoming as real as any building, and just as much a matter to be defined and protected by law. (Though, as Dan Burk's analysis argues, this in itself is a thoroughly complex and problematic metaphor.)[76]

Spam continues to be a very useful problem for thinking through the relationship between law and the Internet, and a full history of spam in law would be a valuable document of how our society struggled to find accord with its networks.[77] Within the United States, the most significant of these struggles was CAN-SPAM. To understand how this legislation came into being and what the stakes were, we must understand how the volunteer antispam movement had grown and how their vision of the network's political role and stakes had evolved.

THE CHARIVARI IN POWER: NANAE

The news.admin.net-abuse hierarchy of conversations on Usenet, the combined police log and Davos Forum for the wizardly classes, was becoming obviously overloaded, its discussions swamped with complaints and the proliferating practices of bad actors online—particularly "unsolicited and/or unwanted mail." Tim Skirvin, then a computer consultant in the Department of Mathematics at the University of Illinois, moderator of the Usenet group humanities.philosophy.objectivism, and author of a central FAQ on cancel messages, submitted a Request for Discussion (RFD), a formal document used to propose changes to the structure of Usenet. It argued for significant reorganization and the creation of some new discussions in the hierarchy, including one called news.admin.net-abuse.email. The proposal was presented on July 9, 1996, and passed by a vote of 451 to 28: "A forum for discussion of possible abuses of e-mail. Possible topics include mailbombing, denial-of-service attacks, 'listserv bombs,' unsolicited and/or unwanted mail, email address lists, mailing list abuse, large-scale mailings in general, chain letters, 'email viruses' such as Good Times, chain letters such as MAKE.MONEY.FAST, filtering software such as procmail, and so forth." The antispam movement now had an official headquarters, acronymically referred to as NANAE.[78]

This was the vanguard party for a coalition government, one that could combine the freelance research-and-punish swarm of the anarchistic charivari, the increasingly sophisticated systems of legal recourse being developed by the parliamentarians, tools adopted from the technolibertarians, and an active enforcement role, where possible, for the wizards at the ISPs and backbone facilities. This coalition gave the antispam movement new social and political structure. It needed the support: the charivari was the expression of powerful and shared system of implicit values, online as in a medieval village, and the network was becoming deeply heterogeneous and ideologically diverse. That post–Canter and Siegel press release drafted at a terminal at the University of Chicago, with all the awkward metaphors for spam, was speaking on behalf of a so-called implicit social contract of the users, but for whom was it speaking in the mid-to-late 1990s? The shareholders of AOL? The millions of new users coming on every year, in businesses and homes and high school computer labs, who had no interest in the endless discussions as to what they could and could not do with their computer and paid connection? Should it reflect the opinions of Ira Magaziner, the Clinton policy advisor who made it very clear in his dealings with Jon Postel that the Internet was the jurisdiction and property of the U.S. government and an engine of commerce?[79] Built on a changing network that could no longer depend on rough consensus and collective voice about its constituency and purpose, without the horizontal technical culture and shared expertise that made Feinler's "'Us' . . . the network users" possible, NANAE provided a common ground for hammering out a new understanding of spam.

NANAE became the central hangout of the antispam community, complete with complex ideological discussions, in-jokes, and a great deal of slang and folklore. By dint of their work tracking new abuses and techniques, they also became the archive of spam in its development, a real-time reference manual distributed over more than a million messages for the intricate, arcane developmental history of "address harvesters," "throwaway dialup," "listwashing," "Joe-jobs," "pink contracts," and so on. At times, the proliferating technical-criminal jargon and the rhythm of the posts, with their alternation of gossip, invective, jokes, and details of network research, resembles the transcripts of some kind of global police precinct. In the process of reporting their activities, they also provided a fragmentary manual for do-it-yourself antispam work, one that transformed with the technical times. "I went a little bit over the top, searching this one out," begins a discussant from Rensselaer Polytechnic Institute (RPI), almost immediately after the formation of NANAE, going to work after receiving one spam message too many from the same group:

```
Here's what I found . . . :

> whois -h rs.internic.net moneyworld.com Financial Connec-
tions, Inc (MONEYWORLD-DOM)

2508 5th Ave, #104 Seattle, WA 98121

Domain Name: MONEYWORLD.COM

Administrative Contact, Technical Contact, Zone Contact,
Billing Contact: Williams, Bob (BW747) d . . .@CYBERSPACE.COM
206 269 0846 ...

> traceroute ns.moneyworld.com traceroute to ns.moneyworld.com
(205.227.174.6), 30 hops max, 40 byte packets

1 ext-gw-2 (128.213.2.1) 3 ms 3 ms 10 ms 2 vccfr1-83.its.rpi.
edu (128.113.83.254) 6 ms 3 ms 3 ms 3 vccfr3.its.rpi.edu
(128.113.100.237) 5 ms 8 ms 5 ms ...
```

```
I'd say that MCI seems to be the culprit, right now, based
on the address in the received column.

Unfortunately, usa1.moneyworld.com and moneyworld.com don't
seem to be related. One's through MCI, the other is
bbnplanet.

I've gotten two e-mails from these guys before. If anyone is
in Bob's area code, could you please give him a phone call
and . . . well . . . play a 9600 baud squalk [sic] in his
ear?[80]
```

This message is an exemplary artifact of the NANAE process. Look at the modes of speech and address that are at work in here: a specialized command, "whois," that requests information from a server about who registered a domain name; a network diagnostic tool, "traceroute," that displays the route and timing of a packet of data to a particular address; the data returned from these commands; a little speculation about culprits; a community request for assistance; and a somewhat tongue-in-cheek suggestion for punishment. On NANAE, system administration tools became the starting points for increasingly elaborate campaigns of surveillance, following the track of messages and the ownership of accounts to trace and identify spammers. (Those who had come after Canter and Siegel had learned from their example not to provide too much personal information in their message—recall Garst changing her phone numbers.)

The earliest problems in NANAE's approach were not technical but social. Once the source of the spam had been identified, early NANAE posts often lapsed back into the confusion of enforcement, punishment, and censure. What exactly were the antispam activists supposed to *do* with the spammers? NANAE came into existence partially because the charivari techniques had failed, in a vague and prankish flurry ("and . . . well . . . play a 9600 baud squalk in his ear?") that left both past and future perpetrators undeterred. The history of which NANAE is both the venue and the transcription is mostly a history of trying to develop enforcement tools, as well as simply keeping track of spam. Angry mail gives way to an elaborate system for formal complaints, and then to the support of lawsuits and legislation, as the exigency of spam's growth forces the development of a citizenry, of sorts: a borderline legal collaboration that could match the inventiveness of their enemies' practices and techniques online.

"I consider myself to be the manager of this FAQ for the good of everyone, not the absolute & controlling Owner Of The FAQ. . . . If the community wants something added or deleted, I will do so."[81] So begins an exemplary document of the NANAE coalition, one of the high-water marks of the social strategy for antispam generally, a single 24,000-word web page (about a quarter of the length of this book) maintained by Ken Hollis, an aerospace engineer at Lockheed and NASA. Hollis is a wizard so wizardly that he goes by the handle "Gandalf the White" and includes in his posting signature the Tolkien quote "Do not meddle in the affairs of wizards, for they are subtle and quick to anger." The document, "alt. spam FAQ or 'Figuring out fake E-Mail & Posts,'" is a garrulous and occasionally repetitious collection of working notes both sent in by others and accumulated by Hollis himself. It includes lengthy descriptions of how to use tools such as whois and traceroute and how to extract the data-laden headers of email to figure out where a given spam message originated; editorials about some of the legislative attempts to curb spam described earlier, with contact information for U.S. elected representatives; a slowly accumulating batch of spam filters; best practices for policing mailing lists; extensive link-laden guides to the then-current range of viruses and malware, come-ons, scams, and fraud; and more. It is a good candidate, among a few somewhat similar others, for the position paper and action guidelines of the vernacular antispam movement. It's full of advice, presented by handle or first name, that is reminiscent of the hands-on, do-it-yourself mood of underground papers of the 1960s and 1970s: "JamBreaker sez: Be sure to let the traceroute go until the traceroute stops after 30 hops or so. A reply of '* * *' doesn't mean that you've got the right destination; it just means that either the gateways don't send ICMP 'time exceeded' messages or that they send them with a TTL (time-to-live) too small to reach you."

One of its most intriguing features is the start of a procedural toolkit for working with governance and laws, a case-by-case and sometimes country-by-country approach for bringing legal authority to antispam activity. There are instructions for contacting the attorney general's office of a U.S. state to check on business licenses and find the names under which they were registered, for instance. On a larger scale, there is a remarkable series of introductions to laws as starting points for debate: "You should also read Title 47 of the United States Code, Section 227.

There is a FAQ at cornell.law.edu for the text of the law"; "Norway—Sylfest tells us Norwegians should report these via email to the national taskforce on economical crime, the KOKRIM by forwarding the mail with full headers to . . ."; "And from the Canadian Department of Justice server . . . STATUTES OF CANADA, C, Competition—PART VI OFFENSES IN RELATION TO COMPETITION—Definition of 'scheme of pyramid selling'—Section 55.1 EXTRACT FROM THE CANADIAN CRIMINAL CODE." Pages like Hollis's—and there were many others along similar lines—promised to make the craft of antispam as transnational and as easy to learn as spamming itself, a roster of techniques with increasing degrees of sophistication in a global framework of borders, states, and laws.

The global network is local at every point, and one of the first items on this page of material is quite geographically specific: "If you are in the United States and have not yet written to your Senator or House of Representatives about how terrible the CAN-SPAM act is, I would ask you to do so." CAN-SPAM was a product of many different legal, political, and market forces, but for the purposes of our analysis, it marked a major transition in the antispam community and was simultaneously the success and failure of the parliamentarian approach.[82] CAN-SPAM's metaphor was simply that spamming was a business like any other. It set aside many of the novel technical implications of the project of spamming and treated spam as local and human in origin—as a group of people with U.S. phone numbers and addresses, engaged in an underregulated form of commerce, who could be easily contacted and regulated conventionally. This was not an unreasonable supposition. The worlds of spam and antispam, at that point in time, could be quite intimate. The most active spammers were a relatively small cohort, based in the United States and known by name, address, and phone number to NANAE and thus to the authorities—and the spammers in turn knew NANAE and followed the conversation.

For the lawmakers and lobbyists forging CAN-SPAM (introduced by future U.S. presidential candidate John McCain), this quality of locality was crucial. From the legislative perspective, the spammers were not so much malefactors as opportunists and entrepreneurs in the wild early phase of an unregulated market. They still fell within the jurisdiction of U.S. law and would respond to it. The means of regulation, cutting clean through the Internet's complex political typology, would not need to take place on the network at all. The network's technical properties were ancillary to its social, geographical dimension, which was the province of the FTC. The long, complex, bitter debates between technolibertarian filter advocates (block offending messages!) and parliamentarian user-educators (train people not to respond!), between wizardly pricing proposals (raise the cost of sending email in bulk!) and vigilante exploit action (blow up their fax machine!), meant nothing here. This new medium needed governmental regulation, and those regulatory strictures would be enforced by the government within the national boundaries on the assets and persons of those responsible for breaking the laws. "That idea is to hold responsible either the person sending the e-mail, or the entity for which the spam is an advertisement," as Lawrence Lessig summed it up.[83] CAN-SPAM was the assertion of one narrow metaphor, and one specific model, of both spam and the network it plagued.

This metaphor was to work by making spam publicly accountable so that the logic of market could handle the heavy lifting of removing unwanted operators while permitting "Internet marketing" as a whole. The law said, in essence: "Spam me in an irritating fashion and I can choose to refuse future mailings, meaning that you lose a potential customer and eventually go out of business," a straightforward market-votes-with-its-dollars understanding of advertising online. The accountability necessary to make spam subject to regulation and market pressures would be produced by three restrictions: the message sender's identity and "from" address had to be authentic and clear so that they could receive replies; the subject line of the message had to indicate that it was an ad; the message had to include text making it clear how to unsubscribe from future messages—regulatory verbiage like "To unsubscribe from these mailings, click here."

The great threat, as seen by antispam activists and many on NANAE, was that this legislation would render spam acceptable. CAN-SPAM was simply the legitimation of spamming for those businesses capable of hiring lobbyists and thereby marking their own messages as reasonable and appropriate. It would allow ISPs or email hosting services to sell their lists of addresses to CAN-SPAM-compliant marketing companies, who could push out streams of messages on behalf of the many large corporations and millions of small businesses in the United States, each of which would need to be opted-out from in turn, and possibly reported for failure to comply.[84] Suddenly, it becomes the responsibility of everyone with an email address to opt out, over and over; direct email advertising is presumed legitimate until proven otherwise. Additionally, the rule of thumb for anyone with a cursory experience of dealing with spam was never to click on "unsubscribe me from this mailing list" links, "because," to quote Paul Graham, a programmer whose arguments about filtering were to change the shape of spamming, "that tells the more unscrupulous spammers that you are a live target who actually read the mail, and you'll just get more spam than ever. Naturally, the opt-in spammers know this."[85] Borderline legitimate and less scrupulous marketers could take advantage of this legislation, leaving the recipient always unsure as to whether they were removing themselves from one list or putting themselves on many, many more.

CAN-SPAM's passage provided the grounds for some major arrests, most notably that of Robert Alan Soloway on charges of fraud, money laundering, and identity theft in 2007 (along with a number of state antispam laws). Although these changed the population and methods of spammers, they didn't slow the growth of spam's volume of messages; indeed, one of the most intriguing things about CAN-SPAM was the technological ductility with which the spammer community responded to it. As far back as the late 1990s, spammers had been including excerpts from proposed antispam legislation and links to legal sites to suggest that they were cognizant of the laws and within their rights. (At one point, Professor David E. Sorkin of John Marshall, who ran spamlaws.com, had to post a notice explaining that he was not responsible for the disclaimers in spam emails linking to his site.)[86] CAN-SPAM provoked similar legitimating moves. It created a market in "valid froms," batches of working email addresses manually created at free accounts all over the world, to satisfy the requirement that advertising messages have a "real address"; at $25 a month for fifty valid from/reply-to addresses, the savings for the spammer was in time by not having to maintain them. There were "CAN-SPAM-compliant" versions of spamming programs that used rented email servers in China, rather than misleading proxy servers, making the source of the message clear (and thereby satisfying the requirement), just as it made the sender's activities far more legally ambiguous and difficult to attack.[87]

The few high-profile busts made possible by CAN-SPAM and U.S. state laws (such as Eliot Spitzer's $20 million suit against Scott Richter in 2003—"We will drive them into bankruptcy, and therefore others will not come into the marketplace to take their place") combined with the boom in filtering systems to rapidly empty out the consciously self-legitimating culture of email marketers.[88] The problem with laws against spam, as Paul Graham put it, was that "the worst class of spammers ignore them"—exactly the sort of people who were going to take control of the business after the laws and the filters changed the spam landscape completely.[89] That then-upcoming pincer movement and the new forms of spam that were to come from it are chronicled in the next section of this book. There were entirely different forms of spam, spun up out of ancient stories and globalized ruin, already germinating outside the United States: "Dear Friend in Christ, I am Marus Joko and My younger sister Mercy is 15yrs, we are from sierra leone but residing in Ivory Coast west Africa."

YOU KNOW THE SITUATION IN AFRICA: NIGERIA AND 419

A poor man dies un-noticed because not many people know him, but when a rich man becomes mere sick, radios, newspapers and other information organs will report it with sensational headlines.

—Sunday Okenwa Olisah, "Life Turns Man Up and Down" (1964)

It was not stem cell research or landing a man on the moon, but packaging a mugu was a science of its own.

—Adaobi Tricia Nwaubani, *I Do Not Come to You By Chance*

A message arrives, a panicked plea making reference to a desperate situation in an exotic location: a wealthy refugee family trying to make it out of Zimbabwe, the widow of an aide to Saddam Hussein in a hospital in Chiang Rai, a Russian oligarch's daughter hiding in the Czech Republic and communicating through her London solicitor. They are looking for a compassionate soul ("whom God will use to assist me and my family") who can help them get themselves and their assets ("US$45,000,000.00 (forty five million united states dollars only)") out of this difficult moment in geography and history. The phone and fax numbers work, the web addresses ("View the above news page for confirmation") point to real news sites—"You can go to google in internet and check my clients name and information, former Governor, James Onanefe Ibori"—and many of the government bureaus and banks check out online. (Christian Eich, who died with the crash of the Concorde in 2000, became a recurring figure in bequest messages—in which the message recipient has inexplicably come up in the will or next-of-kin search—as a sad fictional continuation to his brief obituary in the news.) Within this structure of real locations, events and people is woven an intricate set of recurring fictional themes. There are companies such as Anglogold Corporation, Apex Paying Bank, Liquefied Natural Gas, Petrol Ivoire, Novokuibyshersk Oil—their names as evocative of remote and unknown sites of money and power as some invented Damascene castle or Galician dungeon must have been in the first centuries of the Spanish Prisoner con, which is the *urtext* of what is played out in these emails. You play with real money, your money, and receive real-world artifacts, mostly documents, to pull you deeper into the story. (A raid in Lagos following a politically scandalous scam project that involved forging President Obasanjo's signature found blank money orders and checks and plane tickets, customs documents, passports, university degrees—materials out of which to fashion a seemingly realistic experience.)[90]

This is "419," or advance-fee fraud, and it is already instantly recognizable ("419" is the designation of the Nigerian criminal code referring to fraud and the adopted name for that genre of spam abroad). It is so unmistakable as to have become its own parodic genre. It appears, casually referenced, as a gag in television comedies like *30 Rock* and *The Office*—so often, in fact, that it is the basis for the metagag in which the offer turns out to be legitimate and the "victim" actually makes money, or loses a fortune for their skeptical cynicism, rather than their gullibility. "Scam baiting," the activity of trying to prank and manipulate the spammers, is an elaborate subculture to itself with recognizable tropes and activities such as making the scammers reenact scenes from movies, produce nonsensical artifacts, or travel to dangerous places.[91] 419 has become a cliché, the canonical line ("Hello! I am a Nigerian prince") trotted out in conversation, along with references to Viagra, when describing spam as a category. Because of this quality of cliché, and its consequent facile thinking, we need to bring something very different into this discussion to get proper perspective and to reframe and see 419 as a regional spam phenomenon in depth, with its own peculiarities as a family of stories. These are not ads for products—for porn or mortgages or relief for masculine anxiety—but an enormous narrative about the failures of globalization from which you, the reader, can profit. The messages can be read as a story, first and foremost, the confluence of CNN with a genre that runs back centuries to English confidence tricksters.

The older story, the Spanish Prisoner, in brief: there is a beautiful, rich woman incarcerated by the cruel King of Spain for complex political reasons. You have been contacted because you could help her escape. In return for this, she will give you some part of her fortune (and possibly more). The escape is complex: there need to be bribes for the guards, hired guides and supplies for the trek through the mountains, help for the inside man. You receive pleading notes from her, smuggled out, and letters of credit that will make you wealthy once she and her assets have been reunited. Things do not go smoothly, because Spain is a far-off foreign country in turmoil, politically confusing and corrupt. The official documents are procured, as are the seals and letters of transit, but there's a change of authorities and a new set of bribes is needed. The muleteers have to be paid off. Negotiations have broken down; the prisoner has fallen desperately ill and needs a doctor, which the prison won't provide—but you can help. If you are out of money, could you borrow some? Why lose your savings for the cost of a little additional investment?

It is a persistent con, changing to suit the times and political circumstances. "He fumbled in his pocket and produced the following," writes Arthur Train—the turn-of-the-century author of legal dramas and novels including *Paper Profits*, about Wall Street speculation and the dream of "fortune from nothing" leading to ruin in 1930:

```
MADRID, 7th I—, 19—.

Gentleman: Arrested by bankruptcy I beg your aid in the
recovering of a trunk containing two hundred and fifty
thousand dollars deposited at an English station, being
necessary to come to Spain to leave free the seizure of my
baggage, paying the Tribunal some expenses in order to take
to your charge a valise . . .[92]
```

Train continues quoting this document, so immediately recognizable, just like what I received yesterday with a different geography of crisis, though Train was writing in 1910. "Havana used to be a favorite place," goes a contemporary *New York Times* account of a domestic Spanish Prisoner syndicate operating in the United States in 1898, describing their preferred location of disaster and fortune. "But it is not used now, probably because communication with it is so frequent and easy. The letter is written . . . as fairly well-educated foreigners speak English, with a word misspelled here and there, and an occasional foreign idiom. The writer is always in jail because of some political offense. He always has some large sum of money hid."[93]

The stakes were a bit different then, and the communications have a quaintly Dickensian cast to them. The 1898 syndicate's person in need was a Captain "D. Santiago de Ochoa," having by a convoluted course come to be imprisoned in Cuba, with a trunk full of French banknotes buried in New Jersey, of all places, waiting to be recovered. Exact directions to the trunk are concealed in *another* trunk's false bottom, this last being held in pawn by de Ochoa's daughter's "hard-hearted boarding school mistress for board," a situation that seems more appropriate to Mr. Micawber's misadventures then the international profiteering that usually frames the con. A clipping from a Spanish newspaper is included with the document to verify the facts of de Ochoa's arrest, and the recipient is encouraged to send a telegraph message to the captain's faithful servant in Valencia, Spain.

Consider all the different levels of literary performance at work here across more than a century. Geographically, it plays on the remoteness and romance of a location, a pliably vague background for tropes from fairy tales and popular thrillers such as beautiful, unjustly imprisoned aristocrats and daring rescues. Temporally, it draws strength from current events, particularly war and political chaos—situations in which it is plausible to imagine unlikely fortunes and desperate bargains. It has great narrative pull and traffics in beautiful inscriptions and convincing documents, from that one-off Spanish newspaper clipping in Train's story and the "thin, blue, cross-lined paper, such as is used for foreign letters" mentioned by the *Times*, to the JPEGs of gold bars and the faxes on UN letterhead today, which are presented as proof of the story. Over its full telling, which ends with the exhaustion of the listener's funds, it takes advantage of the human cognitive difficulty called the "sunk cost fallacy"—the amount of money we will spend to recover what we have already spent.[94] This mildly retrofitted story has been seamlessly adapted into the technological platforms and practices around spam. From letters and telegrams in a world of newspapers to email messages in a 24-hour news cycle, 419 works where the spammer's capacity to generate evidence exceeds our individual capacity to evaluate it—given some willful suspension of disbelief.

Not too much suspension is required, though, for reasons anthropologist Daniel Jordan Smith has described: the structure of 419 messages is predicated on a general understanding of the operation of a profoundly corrupt society, and actually reenacts this corrupt operation, exploiting a history of exploitation.[95] This predication is apparent on both the sender and the receiver's sides of the message. From the perspective of the senders of 419 messages, working in Internet cafés at 70-cent-an-hour computers (or $2 for a full night's use, ending with dawn and the resumption of normal business hours), the messages are a natural enough business decision in a society that is, in fact, profoundly corrupt. It is common knowledge that the country's political and business elite actually *do* move millions and even billions of dollars out of the country covertly in collusion with Western business partners and banks—there are plenty of African industrialists and dictators who cut deals with people overseas to send money abroad in return for a kickback. Furthermore, the countries these elites run are so thoroughly corrupt that any significant advancement—any construction of a building, resource extraction project, even getting a phone line or a lease—involves some palm-greasing and "additional costs." If that's the case, how do you expect to make any real money without following their lead? There is a quality of historical tragedy reenacted as farce in these messages, which replicate the kind of deals made by Nigeria's leaders with companies such as Halliburton, Enron, and Shell, and their money laundries and Swiss banks, replayed on behalf of the people who got nothing from those deals.

On the recipient's side, it takes a deeply cynical (if ill-informed) understanding of politics—not necessarily Nigerian, as the messages are often set in other presumably chaotic and corrupt environments, such as Russia or American-occupied Iraq—that views the world as including these covert, corrupt machinations from which you are finally in a position to profit. This cynicism is combined with an almost touching naïveté on the part of the Westerners responding to these messages, not simply in taking it for granted that someone would actually work with them to smuggle millions in gold or launder some huge sum in dollars, but in the apparent fantasy that they would simply then *have* the money, without attracting the attention of Interpol, the Economic and Financial Crimes Commission, the IRS, or the FBI with those multimillion bank transfers or gold shipments. There is a perverse kind of brilliance to the business of 419 messages, which turn the very fact of Nigeria's history of exploitation by

Western interests and its own leadership into a resource that can itself be exploited—as a place in which outsiders can be convinced that they too can take advantage and make a fortune.

But who is actually doing the exploiting? Not, by and large, the writers of the messages themselves, because they are merely fishing for marks, who are then passed up the chain to a smaller group of bosses. The bosses are the kind of people with the resources and expertise for the fax messages, letters, time-stamped photos of gold bars, and so on—and address lists purchased online, which may cost only a few dollars but nonetheless requires a credit card, which the young people in Lagos Internet cafés are quite unlikely to have. Smith, who has lived in Nigeria for a number of years and is married to an Igbo spouse, was able to gain some measure of trust not afforded to a Western journalist, much less law enforcement. He quotes a young 419 writer he interviewed: "How much do I really get from this anyway? The people getting rich from this are the same people at the top who are stealing our money. I am just a struggle-man."[96]

These low-level, somewhat educated scammers, in a society largely without opportunity for those without connections by birth or patronage, have ended up as components in a strange kind of writing machine. This system is made up of young people and old computers telling and retelling stories from templates circulated by email and thumb drives, with names changed and details updated with fresh material from the news. Out of this process, with reliable elements repurposed with new events, another iteration of the story is produced, one among thousands of variants: U.S. soldiers have found a cache of Hussein's gold; a natural-gas oligarch needs to spirit his money out of Putin's Russia. The higher-level bosses, with their stolen or manufactured stationery from Nigeria's U.S. embassy, nongovernmental organization (NGO) offices, and central banks and the money to arrange settings for plausible overseas meetings ("the big store" in the confidence lexicon, a believable space for the con to play out) are drawn from the ranks of white-collar professionals such as attorneys, accountants, and engineers.[97]

Jenna Burrell's ethnographic work among Internet scammers in Accra captures a somewhat similar dynamic of exploiting a legacy of exploitation in a quite different West African context.[98] Her research places the work of advance-fee fraud messages as a subset in the larger project of establishing overseas relationships online for many different purposes—especially the essentially benign and enormously popular Ghanian practice of collecting "pen pals" and creating friendships, romances, and bonds of mentorship or patronage online. This parallel case in another country further emphasizes the ingenious strangeness of the tactic adopted by the messages, with the writers obliged to create stories that model the stereotypes of their region that they think their audience already believe, to confirm what they expect to find, and to trap them. It can be seen, as Burrell suggests, as an operationalized form of W. E. B. Du Bois's "double consciousness"—the obligation to see oneself through the racially charged and stereotyped perceptions of others—which can employ the vague and ignorant sense of Africa as a largely continuous and uniform environment of poverty, war, resource exploitation, and corruption. "You know the situation in Africa," she quotes one of her interviewees, describing his framing of a begging letter: whether seeking pity or manipulable greed, the writers know that they will have better luck with a tale of wartime orphanage and AIDS rather than trying to gather funds to start a business, take IT classes, or move into one's own apartment, which may in fact be the case. Of course this tactic has the effect of further reinforcing the existing stereotypes of Africa generally, both by adding to their retelling as well as producing the more general fear of scams, confidence tricks, and corruption on the part of foreign investors. "It is evil, greedy behaviour and they bring shame to Ghana," Alice Armstrong quotes a Ghanian about the phenomenon of Sakawa witchery—the complex of rumor and anxiety about young men turning to witchcraft practices to bring greater success with online fraud activity.[99] They seek, so the rumors say, control over the minds of foreigners on the computer. The structure of the rumors forms a perfect image of a corrupt and destructive gain, a "diabolical fertility" (one trope describes "Sakawa boys" shape-shifting into snakes that vomit masses of banknotes) that produces short-term gain for a tiny minority while ruining the society as a whole.

The advance-fee fraud strategy was based on a large volume of messages long before it became digital. In the 1980s, after the financial collapse of Nigeria's oil bubble (a bubble that, to be clear, gave even less than bubbles usually do to the general welfare), the scam first became popular as a paper-based business in a complex culture of criminal practices involving counterfeit postage and forged letterheads. When U.S. postal officials cracked down on the mailings in 1998, they seized 2.3 million letters coming through JFK Airport in three months.[100] Though quite unlike spam technically, this wave of postal fraud functioned because of similar economics: the price of mailing was kept artificially low through counterfeiting, and the money to be made from a successful message far outweighed the production costs. Like phishing and related identity-theft activities, a successful 419 message could pay out anywhere from hundreds of dollars, pounds, or euros to hundreds of thousands. With that irregular rate of return and low overhead, there was no reason not to cast a net as wide as possible.

It is difficult to get estimates of replies to 419 messages because few of the successful scams are reported by the victims—bear in mind that they involve both the shame of gullibility and the potentially more serious consequences of the victim agreeing to engage in illegal activities. Periodically, the most catastrophic successes turn up on the public record. In 2000, a businessman named James Adler sued a long list of defendants, starting with the government of Nigeria and the country's central bank, after having paid out $5.6 million over the course of years to participate in transferring stolen funds out of the country, with the promise that he would receive a 40 percent cut. (Unsurprisingly, the court did not carry the case forward, given that Adler was suing to recover bribes and other expenses involved in defrauding the Nigerian government of $140 million of fraudulently over-invoiced funds: "Adler dirtied his hands by intentionally attempting to aid and abet the Nigerian officials' scheme to steal from the government, and by paying bribes.")[101] A Czech retiree named Jiří Pasovský murdered a Nigerian secretary at the country's Prague embassy in 2003 after being told the Consul could not help him recover his losses, which amounted to around $600,000.[102] (This event joins the grim list of 419 scams that end with suicide or murder on the part of the defrauded victim.)

Then there is 419's notorious Everest, starting with a scam built around the promise of an airport construction investment: from 1995 to 1997, the entirely respectable international banker Nelson Sakaguchi beggared Brazil's Banco Noroeste, transferring $242 million in an effort to secure the deal and then to make back and conceal his enormous losses. (Misha Glenny, one of the great journalists of globalized crime, compares the scale of the disaster to Nick Leeson breaking Barings Bank and the looting of the Iraqi National Bank in 2003, to put this example into perspective.)[103] Even if one cannot land one of those successes, a small-time take from students hoping to get out of debt, retirees emptying their pension funds, or wage workers borrowing money from friends still pays out better than many other kinds of spam messages—given that the cost of a thousand messages is basically zero. What makes 419 so remarkable is that all of this business, this international criminal activity and transfer of wealth and the creation of a small population of specialized, almost craftsman-like spammers, is the constantly metamorphosing story of the Spanish Prisoner—possibly, quantitatively, the most told and retold story of the twenty-first century so far.

The international cultural impact on Nigeria has been striking. No other country has become so synonymous with spam, even though, taking spam generally, the vast bulk of the volume has come from the United States and (much less so) from China, Russia, the United Kingdom, and Brazil. As Smith points out, 419 messages have only added to the deep unease outside investors have toward Nigeria, building the perception of a country of thieves. Within the country, "419" has a much broader meaning of general fraud, much of it directed against other Nigerians. 419 can mean the vast frauds perpetrated against the population by political and business leaders working hand-in-glove with foreign corporations such as Royal Dutch Shell (which bragged, in one of the cables made available by WikiLeaks in 2010, of having thoroughly infiltrated the Nigerian government's ministerial staff, knowing "everything that was being done in those ministries") or the small-time work of quack medicines and procedures ("Elekere Agwo: The Quack Doctor," one of the grimmer tales from the notable body of Nigerian literature produced around the Onitsha market, is still sadly contemporary) and everyday scams like selling or renting homes under false pretenses and illegal "tolls" collected by the police.[104]

A subgenre produced by the thriving and astonishingly creative "Nollywood" video industry are 419 pictures, which are devoted to the travails, disasters, and moral turpitude of the scammers who prey on one another and their own people. (It would be anachronistic to call Nollywood production "films"—there is almost none of the apparatus of film recording, reproduction, and projection here. It is a world of videotape, from the earliest Betacam to current digital video [DV] cameras, distributed on VHS and DVD and built from the start on the new technology.) *The Master* is a representative example: the star, actor and comedian Nkem Owoh, also wrote the song "I Go Chop Your Dollar" for the soundtrack—a catchy song whose content is the laudatory thrill of winning at 419, which is "just a game / Everybody dey play am," that everybody plays. (The song's meaning relative to the movie itself is like the celebrations of the joy of gangsterism in Scorsese's *Goodfellas*—it does not end well—but out of context the song was taken as an out-and-out endorsement of fraud, and Owoh has had to repeatedly explain his opposition to 419.) "National Airport na me get am / National Stadium na me build am / President na my sister brother / You be the mugu [the fool, the mark], I be the master / Oyinbo ["white person"] I go chop your dollar, / I go take your money disappear": it's a vision of a world corrupted, end to end.

We can see the whole shape around 419 now, with the production model built on cybercafés and online news cycles in the shadow of a corrupt society—except for one part, the use of Google and other search engines as a corroborating source for producing authenticity. They were operating in the context of a new information environment—a space with Google in it, or over it. And search had its own very particular and very strange spamming problems.

THE ART OF MISDIRECTION

ROBOT-READABILITY

Consider a flower—say, a common marsh marigold, *Caltha palustris*. A human sees a delightful bloom, a solid and shiny yellow, almost enameled against the dark green grasses of a fen, a flower that has brought happiness with its hot primary for centuries; they're mentioned in Shakespeare as "Marybuds" with "golden eyes." A bee, meanwhile, sees something very different: the yellow is merely the edging around a deep splash of violet invisible to human eyes—a color out on the ultraviolet end of the spectrum known as "bee violet." It's a target meant for the creature that can fly into the flower and gather pollen. The marsh marigold exists in two worlds at once. Other flowers may exist in several more. Cultivated roses have one face for gardeners who discern the heritage of a Damascus or a Sunsprite, another for a bee that gathers nectar and pollinates, another for a waxwing that eats the rosehips, and still another for the aphid that attacks the leaves. It has developed to signal to some, and others have learned to read it. It is a single living thing with many faces.

With this concept in mind, consider the increasing space in our lives devoted to things that are "robot-readable," to borrow a term from the designer Matt Jones.[105] Robot-readable media are objects meant primarily for the attention of other objects. The examples ready to hand are still mostly glyphs, such as bar codes, the vivid Day-Glo of chromakey backgrounds, and the gridded, blocky "QR" codes that have recently migrated from the graphic surround of Japan and Europe to the United States (the squares that appear on billboards, products, and clothes for you to scan with your phone to be directed to a website). But there are also classes of physical media—sensor feeds, wireless signals, RFID tags—imperceptible to the human senses. Things translated for our benefit, such as a stream of data rendered for us as a Twitter feed, may be passed over this system, but it occupies a space on the electromagnetic spectrum that only correctly built machines can access.

From this notion comes that of objects and forms meant to be more *difficult* for certain kinds of sensors to pick up. Think of the strange shape of stealth aircraft, faceted like gems with flat faces and hard angles and covered with novel surfacing materials intended to be harder to target with radar—bizarre and counterintuitive design choices meant only to counteract that particular sensor system. Or the compelling fantasy from William Gibson's novel *Zero History*: the "ugliest T-shirt in the world," with a pixelated pattern that somehow makes the wearer disappear from digital surveillance camera feeds.[106] Artist Adam Harvey has been developing camouflage makeup patterns that can disrupt facial recognition systems—the kind of computer vision algorithms that run on security camera footage, as well as photo-sharing services like Facebook's, to pick out and identify human faces in images—by thwarting their particular biometric parameters with interfering shapes and patterns. With this in mind, one begins to see in modern life a dense cloud of symbols and signals meant for devices, sensors, and algorithms.

Two consequences spring from this proliferation. The first is one that Jones and his collaborators have considered at some length: we begin to live in a world of overlapping layers of meaning. Or, rather, we always have—we live in a world of those many-faced flowers, where fruit bats, tree frogs, songbirds, mushrooms, and domesticated dogs are thoroughly engaged in symbolic orders of their own, alongside and throughout ours—but now we also increasingly live with a layer created by things we have made but that we cannot understand unaided.[107] The effects of this layer, like data compression artifacts, peculiarities of mapping software, and the texture of objects rendered through LIDAR and motion sensors, are reflected in the work of artists and in our everyday visual experience. Jones has termed this the "sensor vernacular," and writer and artist James Bridle terms it "the New Aesthetic."[108] ("Advancing technology always brings a new way of seeing," as writer Joanne McNeil puts it.[109])

The second consequence is an oblique reflection of this robot-readable world. It is easy for us to observe things that clearly belong to that category—unreadable to us, like the modular robotic chop of a QR code—but, in fact, steadily larger portions of the world as we see it are becoming robot-readable, as algorithms increase in sophistication. Facial shapes and biometric geometry become readable, as do bodily gaits (your stride, posture, and how you hold your limbs), pieces of music—recognized from a hummed passage of melody or picked out from background sound, trees identified from leaves, and sexual preferences from social graphs. And, of course, text: first the recognition of characters, so a scanned page of some eighteenth-century broadside becomes a searchable, indexable, copy-and-paste-ready digital text file, and then comes semantic processing and analysis of the digitized text. This last form is one of the great technological projects of the twenty-first century, a moon shot or the electrification of the world—the industrialization of text, to extract meaning, connect disparate areas, and, above all, to produce relevance and salience. This is Google's great significance and the most interesting thing into which they have been sluicing their enormous supply of engineering and programming talent and massive investment in hardware: the production of immediate salience from the prodigious ore of available data. The history of spam documented in this book is a history of attention gathering. Spammers target areas where salience has been produced—online communities being accumulations of this kind of engaged, focused attention—because it's there that they can indiscriminately bombard, in search of some accidental interest, that one-in-a-million hit, which is all they need. Google's ranking algorithms, determining the top results for a given search, are a new and more diffuse and abstract form of online community: enormously successful and eminently exploitable.

One last remark about life in a robot-readable world, one whose relevance will grow more pertinent over the course of the rest of the book: it means being always conscious of the current edges of readability, where what is available to our robots (computers, sensors, algorithms) is distinct from what's available to us, with our human brains. This is the underlying principle of CAPTCHAs, for instance, which will play a large role in the third chapter of our history. If you've tried to comment on a blog post, set up a page on a social networking site, or obtain an email address with a web service, you've probably dealt with a CAPTCHA. It is a text entry box, a "submit" button, and an image of a string of text: sometimes a word, often a nonsense jumble of letters and numbers, skewed, warped, compressed, and visually confusing. You type the characters in the image into the submission box—"c7C0ghfg" or "overlooks"—and click "submit." By this means, you have proven your humanity, because a CAPTCHA—which stands for "Completely Automated Public Turing test to tell Computers and Humans Apart"—displays the current point at which human character recognition exceeds that of machines. (It is one of many projects that operate at this borderline, such as the U.S. Post Office's Remote Encoding Center, a facility where human readers figure out those increasingly few handwritten addresses on envelopes that are still incomprehensible to automatic scanners.)[110] If their points of weakness can be found, it's quite possible to trick our robots, like distracting a bloodhound with a scrap of meat or a squirt of anise—giving it the kind of thing it really wants to find, or the kind of thing that ruins its process of searching. The robot can be tricked, and the human reached: this is the essence of search engine spamming.

To understand what happens from now on in spam's history, we need to start with the gradual and momentous transition of the relevance of text into something robot-readable, particularly how that happened with search on the web, and thus how search engine spamming came into being—which means talking about Google.

THE COEVOLUTION OF SEARCH AND SPAM

We can start with the enormous fact of the present: Google's effective dominance in the business of finding relevant information in the global north, and how that dominance is reflected in design and use. Statistics about Google reflect this complex dominance as a mix of very large and very small numbers. As of February 2011, Google was the source of 65–67 percent of all searches in the United States (a relatively steady number, recently), trailed distantly by Yahoo! and Microsoft's Bing search system—which, in fact, handles the queries for Yahoo!—and single percentage points for the coelacanth relics of earlier dot-com struggles such as Ask and AOL.[111] These numbers are misleadingly low as expressions of Google's presence, reflecting the sampling bias of those who would let their use be tracked by the comScore measurement system in return for compensation, akin to the problem of coverage bias in land-line-only telephone surveys.[112] A look at the analytics for websites under my control, and an informal survey of friends and colleagues, uniformly pegs Google in the low-to-mid nineties as the source of searches leading to our websites. The very popular question-and-answer site for programmers Stack Overflow shared a month of their traffic data in 2009: 83 percent of visits to the site came from search engines (rather than people clicking links posted in a discussion, for instance, or typing an address into the address bar manually), and of that traffic, Google produced 3.4 million visits; the second place source, Yahoo!, producing only 9,000. "If every other search engine in the world shut down *tomorrow*," wrote Jeff Atwood, one of the creators of Stack Overflow, "our website's traffic would be effectively unchanged."[113] In many markets, Google is essentially all there is—hundreds of times more popular than its competitors. Google also dominates search abroad, including France, Bulgaria, and Israel—though there are major regional contenders, such as the Russian Yandex, the Korean Naver, and Baidu in China—and remains the intermediary for anyone trying to crack the Anglophone market.

Contrary to the expectations around the collapse of the first Internet investment bubble that the dominant interface of the future would be curatorial, TV-like "push media," search has become the chokepoint through which a vast amount of monetizable attention passes. A great deal of this wealth moves through Google's keyword-based advertising systems: websites can become affiliates of the program, placing ads on their pages that change depending on the content of the page and the keywords used in the search that led to the page. Sites receive a cut based on page views and clickthroughs to the advertiser's site. Those are the big numbers—a lot of users for search, a lot of business to be done, a lot of money to be made in ads and commerce. But all of this activity happens in the context of a small number: The top three results to a Google search get 58 percent of the clicks.[114] Although it is difficult to derive entirely trustworthy numbers from the methodological mix of small-sample laboratory work, network analytics, polls, and logging software—most of which, it should be pointed out, are orchestrated by companies whose business is predicated on promising clients to help them reach that space of the top three—all imply that search, playing an enormous role in Internet use, particularly in commerce and advertising, is largely transacted through a textual space of about one hundred words, the length of the first three Google returns and their excerpts. That is the aperture through which almost all that attention and money flows; this paragraph is more than twice its length.

Search engine optimization (SEO), the name for the business of being in the top three returns, is a thicket of rumor, folklore, sophisticated technical activities, and behavior classified as "spam"—some of it nebulously illegal and some of it sanctioned by its adoption into corporate culture. Many of its tactics are folded into spam as an activity, referred to as "link spam," "comment spam," and "spam blogs," but it remains quite different from "spam" as we've seen it so far on Usenet and email. It marks the further transformation of spam into an activity that can be identified across different technical platforms (email, web pages, blogs). It provides the first examples of what dominates the next chapter—the mechanized semantics of spam created by machines for machines, reaching humans only as a second-order effect. Finally, as a specific set of tactics, with a traceable relationship to the development of search engine algorithms and the economics of advertising on the Internet, SEO gives a new set of moves and countermoves in the technological drama of search. These tactics let us follow spam in action: how a method for gaming search systems becomes "spam" and in turn changes the environment and the systems in which it developed.

At the most abstract level, search engines have three elements from their inception to the present day: a spider, an indexer, and a query handler. The spider moves around the web, collecting data from web pages; these data are passed to the indexer, which builds a searchable repository out of it; this index is parsed by the query handler, which tries to interpret a user's search terms to provide a good fit from the index, in the form of links to relevant web pages. Within those three components, a lot of variation is possible, and it is in these variations that the early search engines struggled for control of the market. A spider can be programmed to move around the web in different ways, whether passing from linked page to linked page or by "random walk" or some more specific system, and can be limited to particular sectors of the web, such as a specific group of addresses or type of file. (Spammers have their own spiders and harvesters, programs that behave like search engine spiders but look only for "@" signs, generally indicative of an email address, and collect the words on either side, passing this back to the spammer's database.) The kind of information a spider looks for on the pages it finds can be specified, as can the way the indexer assembles and arrays the resulting data. Finally, the query handler can be programmed for different kinds of engagement with the data, a choice that often reflects research into usability—will users be more comfortable with a natural language, ask-a-question interface, for instance, rather than just typing in keywords, and will that design interfere with the relevance of the search results? Differences in the design of these components are, at least initially, the major distinguishing factor between competing search providers.

In so-called first-generation search, all three components—spider, indexer, query handler—were skewed toward text rather than links as the criterion of relevance, and spam followed suit. (The generational convention distinguishes major technical and social refinements of search systems from the earliest text searches to Google and beyond.) Spiders moved across the web taking advantage of the structural cues of HTML to gather the presumptive meaning of web pages. HyperText Markup Language (HTML) was the format developed in its most basic form by Timothy Berners-Lee for "marking up" a file of digital text so that a web browser can display it, much as a graphic designer will mark up pages of text so that a printer can put them in the appropriate layout and typographic style. Berners-Lee's markup provided a variety of structural elements that could be used to specify the arrangement and appearance of text on the screen, the placement of illustrations, and so on. It meant that each page had two faces, one robot-readable and the other for human eyes: the source, to be interpreted by the browser, and the displayed page, to be read by the person. For instance, the HTML line `<a href="http://www.google.com">Google</a>` would be rendered in the browser's display as a clickable link—just "Google"—in the default underlined electric-blue typeface.[115]

The spiders that the search engines sent out would go through the HTML source of a page, using the structure of the markup to assess the significance of words with greater or lesser degrees of importance and relevance to a search. A word in a URL (for uniform resource locator, the "address" of the page) or in the first header tag—which is the markup for what the human reader would see as the "title" of the page, as in <h1>My Homepage</h1>—was probably more important than one in the body text of a page and would be rated accordingly in the index. In this respect, spidering HTML duplicated the way in which a person in a library will scan a shelf, pull down a book, and look at the title and subtitle, the author, perhaps the table of contents, when deciding whether to read further. A set of elements called "meta tags" were used in HTML specifically for the benefit of search engine spiders, with keywords listed for the page such that they would be invisible to the human reader but helpful to search indexing. Helpful in theory, anyway: though meta tag elements were popularized by early search engines such as AltaVista and Infoseek, they were so aggressively adopted by spammers that metadata was largely ignored by the turn of the century, with AltaVista abandoning the influence of meta tags on search results in 2002. "In the past," said Jon Glick, then AltaVista's Director of Internet Search Services, "we have indexed the meta keywords tag but have found that the high incidence of keyword repetition and spam made it an unreliable indication of site content and quality."[116] The language is standard public relations–speak, but in this sentence we can see that spam has completely shifted from its mid-1990s technical frame. "Spam" here is a noun for the type of language being used in <meta> tags, rather than repetition and irrelevant ranting in a Usenet discussion, or an inappropriate commercial email. In a paper from 1998 devoted to modeling search algorithms, the authors wrote: "Some authors have an interest in their page rating well for a great many types of query indeed—spamming has come to the web."[117] "Spam" has become a general term for an action taken with a networked computing protocol to capture the attention of the largest possible number of people. The noun "spam," with its technolinguistic residue in <meta> tags and "keyword stuffing" on pages, can now also mean a repetitive mass of words generated as part of machine processes, like the frass produced by the activity of termites.

What precisely was the business plan of early search spammers, and what were they putting in their web pages? Keywords offering a promising rate of incoming clicks were repeated in the meta tags and gathered en masse in the page itself, hidden from the casual human reader's eye. One of the details that HTML can specify when you are building a web page is the color of text, so the page's author could set the page's background to gray and make text the same shade of gray, invisible on the human reader's display while appearing to be normal text in the body of the page as far as the spider was concerned. Hence the phenomenon of innocuous pages with some form of spammy intent, perhaps a product to sell, having at the bottom of the page an odd gap. The text on the page has ended and there are no images, just a few inches of the gray background before the bottom of the rendered page. In that gap, in background-matching color and often minuscule font size, lay a magma flow of obscenity and pornography, product names, pop stars, distinctive phrases, cities and countries, and odd terms seemingly plucked from Tristan Tzara's hat, selected because they happened to get good returns at that time. Reading these hidden word sets provides something of the same feeling of lexical torque we will encounter in algorithmically generated "literary spam"—as though a Céline character worked for *Entertainment Tonight:* toyota ireland ladyboy microsoft windows hentai pulp fiction slut nirvana. It is a bizarre reading experience, not meant for our eyes, like walking into a field of flowers and seeing them as bright targets of bee violet and bee purple—messages for another order.

Aside from the feeling of having gained access, by way of the metrics of search returns, to some repulsive stock ticker for the reptilian hindbrain, such blocks of text also illustrate a recurring theme in the development of spam: a matter-of-fact distinction between humans and machines, with different strategies for dealing with each. As spam becomes more sophisticated, more technically complex, and demanding of greater skill to produce, this distinction becomes more prominent. Almost every piece of spam, whether over email or in the context of spam blogs or comment spam, will be *biface*, capable of being read in two ways with very different messages for the algorithm and for the human.[118] As a practical concern, spam explores exactly those technologically and philosophically interesting spaces where distinctions between human and machine activity can be exploited, such as natural language inference, modes of reading, and visual and auditory recognition—understanding the import of pictures, or recognizing text that is visually distorted or spoken, whether in filter-beating literary spam messages or visually distorted CAPTCHAs used to distinguish people from programs. The messenger will not get it, but the person on the other side will understand. Search engine spamming is one of the great proving grounds for the melding of robot and human readability, with different meanings for algorithms and for eyes.

An initial sign of the increasing sophistication in creating biface texts comes from the first-generation search spam technique of "cloaking," which persists in elaborated form to this day. Search engine spiders identify themselves by the way in which they request a web page. This identification is part of the set of protocols that help to distinguish a normal web browser from other platforms, such as a mobile device or an assistive Braille display for the visually impaired. These protocols make it possible to serve a modified version of the page's content to different platforms so that the mobile device will receive a page in a more compact format and a device for the visually handicapped will get alternative text substituted for images. This identification makes it possible to serve one page to a spider, to be indexed and delivered as a search result, and an entirely different page to the user who clicks on the link.

Robotic reading has several unique routines, such as checking for a particular file called "robots.txt," sending the page request from a certain range of addresses, and never loading images. These signatures, which trigger the cloak page, have proved very difficult to disguise; communally maintained sites exist solely to track the addresses known to belong to search engine spiders, keeping logs of monotonous numbers with the same obsessive precision of amateur satellite spotters or shortwave radio hobbyists tracking "numbers stations." For Google's spiders, for instance:

```
216.239.59.99

216.33.229.163

64.233.173.193

64.233.173.194

64.233.173.195
```

and so on. Such lists are largely built out of collective support, as individual website owners note addresses on the access logs and send them in to be confirmed as spiders. A similar degree of labor and ingenuity goes into the business of tracking and managing the use of the cloak pages and real pages, part of the strange business of keeping separate materials for human and nonhuman readers, maintaining a network version of the apocryphal Potemkin village.[119]

At the moment the biface text in its various forms was being developed on spammy web pages, "spam" had only recently made the semantic transition from referring specifically to misbehavior in discussions and on Usenet newsgroups, to email, and then to websites and search engine returns. This territory we have been discussing so far is first-generation search, which is built rather crudely, in retrospect, on simply counting and collecting keywords, like a simple method for finding a library book or a pertinent paragraph in a document. Second-generation search, using "votes" from the link graph, was already underway, however, and third-generation search was beginning to take shape. In 1998, the same year as the paper quoted previously applied the term "spam" to sites, a paper articulated this coming third generation: "Anatomy of a Large-Scale Hypertextual Web Search Engine," presenting "Google, a prototype of a large-scale search engine which makes heavy use of the structure present in hypertext."[120] It is not simply in retrospect, after Google's enormous success, that it stands as a very interesting paper.[121] Studying the "Anatomy" for itself, as an intellectual project, best expresses what it meant for the idea of search and thus the practice of spam—spam being one of the problems with search that Google was originally intended to address.

The Google presented in their paper offers solutions for some of the issues that beset search, such as scalability (the capacity to expand, both in hardware and software, with the size of the web as its numbers exploded—a daunting task) and speed. However, the major conceptual step forward was *social*. Google was not unique in this argument—already relatively simple second-generation search systems such as Inktomi were counting links in weighting their pages—but Brin and Page had a larger plan for search. The "Anatomy" makes a straightforward argument: "The citation (link) graph of the web is an important resource that has largely gone unused in existing web search engines. We have created maps containing as many as 518 million of these hyperlinks, a significant sample of the total. These maps allow rapid calculation of a web page's 'PageRank,' an objective measure of its citation importance that corresponds well with people's subjective idea of importance. Because of this correspondence, PageRank is an excellent way to prioritize the results of web keyword searches." What we see here is the quantification of attention: building a system that can extract the filtering effect noticed by early online community members and turn it into an optimizable component of the search process.

In other words, Brin and Page took the social side effect of a technical hypertext architecture (I link to your page because it is somehow relevant to what I'm talking about or what I want to show my readers) and folded it back into the system, treating it like any other form of data. There were other theoretical approaches to this idea around the same time, all with essentially the same model: take the textual search that is already present in the first generation and build this social dimension into it. The spiders of earlier search engines used the structural cues of HTML to determine the relative importance of words. The spiders of the second generation expanded this approach slightly, using the telltale structure of links on a page to log which pages were linked to and then treating the outgoing links as votes for the linked pages. Google built on these earlier methods: a page is present in the index as a set of "hits," "a list of occurrences of a particular word in a particular document including position, font, and capitalization information"—as it would be in a first-generation search engine's index, with words weighted by their presence in headers, in meta tags, and in the body text and by their proximity to one another. Every hit list, the extracted structure of a web page, is then weighted by the raw count of links to it. Brin and Page then went further: inspired by academic citation structure, they argued for reputation, essentially treating links as a measurable expression of social value.[122]

Part of what this system was meant to address was spam—though they never use the word, which was still making its lexical move across the categories, calling it "junk results" instead. The metric of success Brin and Page set for themselves is "relevance," and one of the key problems with prior first- and second-generation search engines was the increasing interference of spam activity with search returns: "Automated search engines that rely on keyword matching usually return too many low quality matches. To make matters worse, some advertisers attempt to gain people's attention by taking measures meant to mislead automated search engines. . . . Anyone who has used a search engine recently can readily testify that the completeness of the index is not the only factor in the quality of search results. 'Junk results' often wash out any results that a user is interested in."[123] A closely related paper from the following year states outright what Brin and Page allude to: "Linkage on the Web represents an implicit endorsement of the document pointed to. . . . Several systems—e.g., HITS, Google, and Clever—recognize and exploit this fact for Web search . . . because, unlike text-only ranking functions, linkage statistics are relatively harder to 'spam.'"[124] Spam is present as the identifier of an activity, but it is still in the relative safety of quotation marks, the cocoon of a neologism in its larval stage. And it is only "relatively harder": so-called link farms—pages of nothing but links between spam sites—were already doing to second-generation search indices what masses of keywords had done to the first generation, producing an illusion of significance.

How was Google's reputational system, the third generation of search built atop linkage statistics, supposed to stop the keyword stuffers and the link farmers? Spam pages happen to be lonely because a human user can recognize them as such and has no reason to link to them. The only links to spam pages, as a rule, come from other spam pages. Links, in theory, carry an implicit endorsement—a vote of relevance made by a person, like a recommendation made in a community. The spam-fighting question is: who is the person, and how much does their endorsement count for? This is represented as an equation:

> We assume page A has pages T1 . . . Tn which point to it (i.e., are citations). The parameter d is a damping factor which can be set between 0 and 1. We usually set d to 0.85. . . . Also C(A) is defined as the number of links going out of page A. The PageRank of a page A is given as follows:
>
> $$PR(A) = (1 - d) + d\,(PR(T1)/C(T1) + \ldots + PR(Tn)/C(Tn))$$

This straightforward equation is something stranger than it first appears. Combined with a sufficiently large database of pages and links, like the web, it constitutes "a model of user behavior." The model user in this case is the web's version of *l'homme moyen*, the average man or the "random surfer" who performs a version of the mathematical random walk along the link graph of the web. She or he is "given a web page at random and keeps clicking on links, never hitting 'back,' but eventually gets bored and starts on another random page. The probability that the random surfer visits a page is its PageRank. And, the *d* damping factor is the probability at each page the 'random surfer' will get bored and request another random page."[125] It is a rather depressing *l'homme moyen* for our era, having no particular interests while retaining the capacity to be bored, and content to follow links and occasionally request pages at random, always moving forward along the graph.

The *PR(A)/C(A)* element, feeding into the PageRank of linked pages, is what brings a new social element to the process: the economy of reputations. A crude academic citation web will just show nodes and lines, a mass of more and less prolific careers, more and less foundational papers and books whose citations of other documents—links—tie people together. But to the human reader, a paper cited by a Paul Erdős or a Linda Buck or published in *Nature*, in *Critical Inquiry*, or under the aegis of a major university press inherently carries more weight. There is a flow of reputations, as a celebrated publishing machine such as Slavoj Žižek helps to bring the attention of a new market to someone already established in another community, such as Alain Badiou. PageRank thus includes "recursively propagating weights": all links are not of equal value, and reputation spills outward like water in the cups of an ornamental fountain. "A page can have a high PageRank if there are many pages that point to it, or if there are some pages that point to it and have a high PageRank. Intuitively, pages that are well cited from many places around the web are worth looking at. Also, pages that have perhaps only one citation from something like the Yahoo! homepage [that is, from a major, reliable human-curated directory, as Yahoo! was at that time] are also generally worth looking at."[126]

PageRank, in taking advantage of social factors that are "relatively harder to 'spam,'" included one more thing: *d*, the "damping factor." "This allows for personalization and can make it nearly impossible to deliberately mislead the system in order to get a higher ranking."[127] The damping factor is underappreciated as an antispam strategy. It's the subtle gradation of how rank passes through links, how far reputation can go before its effect decays into nothing. The source of the damping factor models is boredom. If a site's pure discoverability to our "random surfer" is the number of inbound links, and if those links are then weighted by the rank of the linking sites—the quantification of reputation—then the possibility that the surfer moving along the link graph will become bored and request a random page is the damping factor. In theory, adjustments to the potential for boredom mean that the flow of importance can be artificially constrained, ensuring that the loneliness of spam pages will make it more difficult for them to siphon relevance from a page linked to by a page with excellent PageRank. If A, with a high PageRank, links to B, and B to C, and C is an easier site to get linked from than A, it will not make much difference if a group of spam pages get links from C; the rank will have diffused too much by then to raise their calculated relevance significantly. You would loan money to your friend A, and maybe to A's friend B, but would you loan it to A's friend B's friend C?

The importance of that idea to Google's prominence remains central (though, of course, they were not the only ones to have it). It is the mark of third-generation search: taking the social side effect of the link graph and folding it back into the weighting of words on pages. This approach was the solution to the problem of relevance, because the matter or the mass of citation and reputation would provide a socially verified evaluation of every page for the system to infer—an unintended communal act of attention. The entire population of creators on the web, everyone in a position to put a link on a page, was voting without realizing it, and their aggregated votes, weight-adjusted for their individual reputations, are a description of what the most-approved page would be for any given set of words. By bringing together the technical-objective and the social-subjective, Google could eliminate the "junk results," the spam that could fool machines but not humans with its stack of keywords. Building on the algorithmic inference of social data, they could make it "nearly impossible to deliberately mislead the system." The only workaround for spammers would be to build their own artificial societies—which is, in fact, exactly what they did.

3 THE VICTIM CLOUD: 2003–2010

FILTERING: SCIENTISTS AND HACKERS

MAKING SPAM SCIENTIFIC, PART I

We have machines capable of reading, analyzing, judging any written text. But it is precisely the reliability of the instruments on which we must run some checks.

—Italo Calvino, *If on a Winter's Night a Traveler*

If you can't get to grips with the spammers using law, censure, and protest, can you instead get to grips with spam itself? How do you get a handle on it, and how do you make it something you can measure and quantify, talk about coherently, understand—and therefore attack? It is an etymologically restless thing, at once noun and verb, that thrives wherever we have trouble defining it clearly enough to exclude it and make precise rules and laws for it. It is the subjective nemesis of the equally subjective germane human interaction. How do you turn it into a material you can work on? How do you make spam an object?

These are matters of practical interest for those seeking to analyze spam and make it something they can stop. This is a story about two such groups—scientists and hackers—and how they went about drawing lines around spam, defining edges and black boxes and criteria and workflows and making it into something to which they could apply tools. This is also necessarily a story about everything left out of the lines they drew, and how in changing the shape of spam it eluded them and transformed, both as a technology and as a set of practices, into something far stranger than before: a new object with a new infrastructure behind it, produced by a new class of criminals.

Spam comes into a computer lab with as much of a halo of strangeness as a chunk of cavorite—H. G. Wells's fantasy material that resists gravity, with which scientists fall up to the moon—and with similarly strange and innovation-demanding effects. After all, what is this human-machine, innovative-criminal, social-technological, maddening yet unstoppable thing? It's a practice, and a communally expressed attitude, but also an artifact of sorts, something that exists in the singular and ostensive—a "spam message," this spam—but that also demands analysis in the plural as *spam*, the problem, on a larger scale. How do you specify this concept, making it productive of reproducible and falsifiable results that are capable of being benchmarked and tested? Spam is constantly fluctuating; the amount you receive depends on your ISP, what filters your ISP uses, your operating system and mail application, the number and type of the accounts that you use, and even the season and the time of day.[1] Spam may seem at first like an ideal subject for scientific testing, as you do not even need to go to the trouble of collecting fruit flies or putting up a telescope to get material—just create an email account and watch it roll in! The infrastructure of spam is so complex, however, that simply testing the result of any given email account is like proving a cure in medieval medicine: the patient may improve or decline, but it is hard to definitively link that change to what the doctor did. Clearly the necessary starting point is some kind of agreed-upon spam object, something like a spam meter or spam calorie, on which and against which things can be tested—a corpus.

A spam corpus, though, starts with the problem of privacy. Simply having a batch of agreed-upon spam messages is not good enough, because spam makes sense only in context, as something distinct from legitimate mail. If your ultimate goal is to produce something that can stop spam, you need an accurate simulation of the inbound email for a user or a group of users in which that spam is embedded. You need not just the spam but also its context of nonspam (referred to in much of the scientific literature, with a straight face, as "ham"). Spam filters improve based on increased amounts of data available for them to work with, so you need a lot of spam, and correspondingly a whole lot of legitimate mail, to get the law of large numbers on your side. Obviously, email is largely a private medium, or is at least treated like one. How do you create an accurate spam corpus, with the vitally important contextual nonspam mail, while maintaining privacy? Simply obfuscating personal details (email addresses, telephone numbers, proper names) seriously interferes with the accuracy of the results. Analyzing email and spam is all about quantifying and studying messages and words in messages, so having a corpus full of "Dear XXXXX" and "you can reach me at XXX-XXXX" is going to make any resulting tests inaccurate at best and misleading at worst for a filter meant for use outside the lab.

What if you obfuscated the entire body of every message in the corpus? You could have a huge set of personal messages and spam without infringing on anyone's privacy. This is a substitution method: "to release benchmarks each consisting of messages received by a particular user, after replacing each token by a unique number in all the messages. The mapping between tokens and numbers is not released, making it extremely difficult to recover the original messages, other than perhaps common words and phrases therein."[2] A *token* is a term from lexical analysis for its basic unit: the atomic part of a document, usually but not necessarily a word. *Tokenization* is what you do to a document to make it into an object for computational lexical analysis: turning the strings of characters we recognize as readers into a series of discrete objects that possess values and can be acted on by a computer algorithm (a "value" in this case being, for instance, the number of times a word appears in the text as a whole). In tokenizing texts, the human meaning of a word is pretty much irrelevant, as its value as a token in a space of other tokens is all that matters: how many there are, whether they often appear in relation to certain other tokens, and so on. It's not as strange as it sounds, then, to preserve privacy in the spam corpus by the Borgesian strategy of substituting a unique number for each token in a message: 42187 for "Dear," 8472 for "you," and so on.[3] If you keep consistency, it could work.

However: "the loss of the original tokens still imposes restrictions; for example, it is impossible to experiment with different tokenizers." Making an obfuscated corpus with this one map of numbers (which is not released, to prevent people from reversing the obfuscation and reading the private mail) locks all users into one version of the corpus, preventing them from trying different methods on the original messages. There are many ways of performing lexical analysis, evaluation, and parsing that could lead to otherwise-missed implications and innovations for further classification and filtering experiments. Starting from a natural language corpus of spam and nonspam messages provides a lot more experimental room to move, compared to a huge pile of integers representing messages already preprocessed in particular ways. ("42187 64619 87316 73140 . . .").[4]

There were other approaches to building a spam corpus for scientific research, but each had its own flaws. A series of corpora was made from mailing lists, such as the Ling-Spam corpus, that collected the messages sent to a mailing list for the academic linguistics community—a moderated list on which someone reviewed and approved each message before it went out, which was thus a mailing list free of spam. In theory, this corpus could be used to establish benchmarks for legitimate nonspam email. In practice, the material appearing on the list was far more topic-specific than the profile of any given person's actual email—no receipts from ecommerce sites, no love letters with erotic language, no brief appointment-making back-and-forth messages. It produced over-optimistic results from classifying and filtering programs for their ability to recognize legitimate text. (Spam messages on the one side and arguments about Chomskyan grammar and linguistic recursion on the other makes for a rather skewed arrangement.) The SpamAssassin corpus, gathered to test the spam filter of the same name, used posts collected from public mailing lists and emails donated by volunteers. It ran into exactly the opposite problem, with a benchmark set of legitimate text that was far more diverse than that of any given person's email account, and also used only those messages considered acceptable for public display by their recipients.

Early in the literature, the solution was quick and rough: the researchers would simply volunteer their own email inboxes for the experiment without releasing them as a corpus. They would hope that the results they got with their experiments could be reproduced with other people's corpora, as do scientists using their own bodies as the experimental subjects—Pierre Curie generating a lesion on his own arm with radium, or Johann Wilhelm Ritter attaching the poles of a battery to his tongue.[5] We can call this the "we're all pretty much alike" approach. "As far as I know," writes Jason D. M. Rennie in 2000 of his early email filter and classifier ifile, originally released in 1996 and described later in this chapter, "there are no freely available data sets for mail filtering. . . . I asked for volunteers who would be willing to have such experiments performed on their mail collection; four users (including the author) volunteered."[6] It's a striking idea, and in many ways one that speaks more to the hacker sensibility than to that of the institutionalized sciences: the code runs, and it's free, so do the experiments yourself. (It would be wonderful for an article in a journal on genomics, for instance, to say that substantive results depend on you, the reader, to fire up a sequencer and a mass spectrometer and perform the experiments yourself—and on yourself, no less.)

Obfuscation, problematic sampling and harvesting, noble volunteers for one-shot tests: the lack of "freely available data sets for mail filtering" that accurately approximate the mailing dynamics of persons and groups was clearly the problem to beat in making spam scientific. Beyond the institutional and methodological issues created by everyone having yardsticks of different lengths, there was a more subtle concern, one native to the task of filtering email and separating it into spam and nonspam. "The literature suggests that the variation in performance between different users varies much more than the variation between different classification algorithms."[7] There are distinctive patterns and topologies to users, their folders, the network they inhabit, and other behavior. We are not all pretty much alike, and acting as though all email corpora are created equal is like a chemist deciding to just ignore temperature. Working with an experimental base of small sets of volunteered email makes it difficult to assess differences between filters and almost impossible to analyze differences in the email activity profiles of persons and groups—differences any filter would have to take into account. Creating a scientific object with which to study spam appeared to be at an impasse.

In 2003, the United States Federal Energy Regulatory Commission (FERC), as part of its major investigation of price manipulation in energy markets, made public all of the data it had accumulated from the energy-trading company Enron. Initially lauded as an innovative corporate player in the energy business, Enron's very public descent into bankruptcy amid revelations of price fixing and massive accounting fraud was the dominant story in business journalism in late 2001 and 2002. The Enron bankruptcy case was one of the most complex in U.S. history, and its related investigations produced an extraordinary quantity of data: the FERC's Enron collection includes audio files of trading floor calls, extracts from internal corporate databases, 150,000 scanned documents, and a large portion of Enron's managerial internal email—all available to the public (at varying degrees of difficulty).[8] The FERC had thus unintentionally produced a remarkable object: the public and private mailing activities of 158 people in the upper echelons of a major corporation, frozen in place like the ruins of Pompeii for future researchers. They were not slow in coming.[9]

The Enron collection is a remarkable and slightly terrifying thing, an artifact of the interplay of public and private in email. As a human document, it has the skeleton of a great, if pathetic, novel: a saga of nepotism, venality, arrogant posturing, office politics, stock deals, wedding contractors, and Texas strip clubs, played out over hundreds of thousands of messages. The human reader discerns a narrative built around two families, one biological and tied by blood, and the other a corporate elite tied by money, coming to ruin and allying their fortunes to the Bush/Cheney campaign. These are the veins of narrative interest embedded in the monotony of a large business. A sample line at random, from "entex transition," December 14, 1999, 00:13, with the odd spacing and lowercase text left in as artifacts of the data extraction process: "howard will continue with his lead responsibilites [sic] within the group and be available for questions or as a backup, if necessary (thanks howard for all your hard work on the account this year !)." This relentless collection of mundane minutiae and occasional legally actionable evidence, stretching out as vast and trackless as the Gobi, was dramatically transformed as it became an object suitable for the scientific analysis of spam.

After the body of the Enron data has been secured in a researcher's computer like the sperm whale to the side of the *Pequod*, it must be flensed into useful parts. First, the monstrous dataset of 619,446 messages belonging to 158 users has to be "cleaned," removing folders and duplicate messages that are artifacts of the mail system itself and not representative of how humans would classify and sort their mail. This step cuts the number of messages to 200,399, which is still an enormous body of text. The collected data for each user is sorted chronologically and split in half to create separate sets of training and testing materials for the machines. The text is tokenized with attention to different types of data within the set—"unstructured text," areas such as Subject and Body with free natural language, "categorical text," well-defined fields such as "To:" and "From:", and numerical data such as message size, number of recipients, and character counts.

Then the testing and evaluation starts, with its own complex conceptual negotiations to make this process scientific. Machine learning programs are run against the training and testing sets and their output is analyzed; these results are then compared to the results from a dataset of messages volunteered from students and faculty at Carnegie Mellon.[10] With a dataset that is cleaned and parsed for computational processing, and not too far afield from the results of volunteer data, the creation of a corpus as an epistemic object appropriate for scientific inquiry is almost complete. This process raises a further question: has the work of making the corpus, particularly resolving what is considered as spam, changed it in ways that need to be taken into account experimentally? The question of what spam is, and for whom, becomes an area of community negotiation for the scientists as it was for the antispam charivari and free speech activists, for the network engineers and system administrators, and for the lawyers and the legislators.

Cormack and Lynam, authors of the recent dominant framework for a spam filtering corpus, directly address this issue of creating "repeatable (i.e., controlled and statistically valid) results" for spam analysis with an email corpus.[11] In the process of establishing a "gold standard" for filters—the gold standard itself is an interesting artifact of iterating computational processing and human judgment—they introduce yet another refinement in the struggle to define spam: "We define spam to be 'Unsolicited, unwanted email that was sent indiscriminately, directly or indirectly, by a sender having no current relationship with the recipient.'"[12] With this definition and an array of filtering tools, they turn to the Enron corpus that we have looked at as processed and distributed by Carnegie Mellon: "We found it very difficult to adjudicate many messages because it was difficult to glean the relationship between the sender and the receiver. In particular, we found a preponderance of sports betting pool announcements, stock market tips, and religious bulk mail that was adjudicated as spam but in hindsight we suspect was not. We found advertising from vendors whose relationship with the recipient we found tenuous."[13] In other words, along with some textual problems—important parameters for testing a spam filter that were missing, such as headers to establish relationships, and the presence of attachments—the corpus reflected the ongoing problem of the very definition of spam. The email culture of academic computer science is not awash in religious bulk mail and stock tips, except insofar as those messages—especially the latter—arrive as spam. But the email culture of the upper echelons of a major corporation is quite different, particularly when that corporation is, as Enron was, both strongly Christian and strongly Texan. As the company began to fall apart, interoffice mail in the FERC dataset includes many injunctions to pray, and promises to pray for one another and other professions of faith. Similarly, sports, and betting on sports, are part of the conversation, as one might expect from a group of highly competitive businessmen in Houston.

In theory, a sufficiently advanced and trained spam filter would algorithmically recognize these distinctions and be personally accurate within the Enron corporate culture as it would for an academic, learning to stop the unsolicited stock tips for the latter while delivering the endless tide of calls for papers. However, within the Enron corpus as tested by Lynam and Cormack, the details critical for such a filter were missing. Human decisions about what qualified as spam, combined with technical constraints that made it hard to map relationships—and "a sender having no current relationship with the recipient" is one of their spam criteria, as it's the difference between a stock tip from a boiler room spam business somewhere and one from Ken Lay down the hall—had made the corpus into an object meaningful for one kind of work but not another. It was an appropriate object for the study of automated classification, but not for spam filtering. In the end, they retrieved the original database from the FERC and essentially started over. "Construction of a gold standard for the Enron Corpus, and the tools to facilitate that construction, remains a work in progress," they write, but: "We believe that . . . the Enron Corpus will form the basis of a larger, more representative public spam corpus than currently exists."[14]

This is all reliable, well-cited, iterative, inscription-driven science: proper, and *slow.* Compared to the pace of spam, it's very slow indeed. A faint air of tragedy hangs over some of the earlier documents in the scientific community when they explain why spam is a problem: "To show the growing magnitude of the junk E-mail problem, these 222 messages contained 45 messages (over 20% of the incoming mail) which were later deemed to be junk by the user."[15] That percentage, on a given day as of this writing, has tripled or quadrupled on the far side of the various email filters. A certain amount of impatience with formal, scientific antispam progress is understandable.

As it happens, networked computers host a large and thriving technical subculture self-defined by its impatience with procedural niceties, its resistance to institutionalization, its ad hoc and improvisational style, and its desire for speed and working-in-practice or "rough consensus and running code": the hackers. They have their own ideas, and many of them, about how to deal with spam. The most significant of these ideas, one that will mesh computer science, mathematics, hacking, and eventually literature around the task of objectifying spam, began with a hacker's failure to accurately cite a scientist.

MAKING SPAM HACKABLE

I thought spam robots would become more sophisticated—the robots would fight antirobots and antiantirobots and eventually all the spam robots would become lawyers, and take over the world. But the filters were pretty good, so that prediction was wrong.

—Robert Laughlin

"Norbert Wiener said if you compete with slaves you become a slave, and there is something similarly degrading about competing with spammers," wrote Paul Graham, a prominent programmer in the Lisp language and, as a success of the first dot-com bubble, one of the founders of the venture capital firm Y Combinator. His landmark essay "A Plan for Spam" is one of the most influential documents in the history of the anti-email-spam movement.[16] The project Graham started by proposing a new form of filtering, which was rapidly taken up by many hands, is important for two reasons: first, because he won—the idea he popularized, in accidental concert with U.S. laws, effectively destroyed email spamming as it then existed, and it did so by sidestepping the social complexities and nuances spam exploited and attacking it on a simple and effective technical point. Second, because he lost, after a fashion: his pure and elegant technical attack was based on a new set of assumptions about what spam was and how spammers worked, and email spammers took advantage of those assumptions, transforming their trade and developing many of the characteristics that shape it today.

"A Plan for Spam," the essay that launched a thousand programming projects, is a series of tactics under the banner of a strategy. The tactics include an economic rationale for antispam filters, a filter based on measuring probabilities, a trial-and-error approach to mathematics, and a hacker's understanding that others will take the system he proposes and train and modify it for themselves—that there can be no general understanding or "gold standard" for spam as the scientists were seeking but only specific cases for particular individuals. The overarching strategy is to transfer the labor of reading and classifying spam from humans to machines—which is where Norbert Wiener comes in.

Wiener, a mathematician and polymath who coined the term "cybernetics" as we currently understand it, found much to concern him in the tight coupling of control and communication between humans and machines in the middle of the twentieth century. He worried about the delegation of control over nuclear weapons to game theory and electronic computers and about the feedback from the machines to the humans and to human society. In the 1948 introduction to his book *Cybernetics*, Wiener made a statement that he would return to intermittently in later studies such as the 1950s *The Human Use of Human Beings*: "[Automation and cybernetic efficiencies] gives the human race a new and most effective collection of mechanical slaves to perform its labor. Such mechanical labor has most of the economic properties of slave labor, although, unlike slave labor, it does not involve the direct demoralizing effects of human cruelty. However, any labor that accepts the conditions of competition with slave labor accepts the conditions of slave labor, and is essentially slave labor. The key word of this statement is *competition*."[17]

For Wiener, to compete with slaves is to become, in some sense, a slave, like Soviet workers competing with impossible Stakhanovite goals during the Second Five-Year Plan. Graham paraphrases Wiener because he wants to present his feelings about spam as a parallel case: "One great advantage of the statistical approach is that you don't have to read so many spams. Over the past six months, I've read literally thousands of spams, and it is really kind of demoralizing. . . . To recognize individual spam features you have to try to get into the mind of the spammer, and frankly I want to spend as little time inside the minds of spammers as possible."[18] It's a complaint that will recur with other researchers—to really understand the spamming process, you have to run an emulation of the spammer in your mind and in your code, and the sleaziness that entails is degrading.

The analogy with Wiener is inexact in an informative way, though, which describes Graham's strategy in a nutshell. Graham does not have to deal with spammers by competing with them. He is not sending out spam in turn or trying to take advantage of their credulity. In no way is he being demoted to the economic status of a spammer by his work, because he is not competing with them—his *machine* is competing with them. He is doing exactly what Wiener predicted, though he does not say as much: he is building a system in which the spammers will be obliged to compete with machines, with mechanical readers that filter and discard with unrelenting, inhuman attention, persistence, and acuity. With his mechanical slaves, he will in turn make the business of spamming into slavery and thus unrewarding. As Wiener feared automation would end the economic and political basis for a stable, social-democratic society ("based on human values other than buying or selling," as he put it), Graham means to end the promise of profit and benefit for small effort that spam initially offers.

The means of ending that promise of profit is Graham's technological tactic: filtering spam by adopting a system based on naïve Bayesian statistical analysis (more about this in a moment). The one thing spammers cannot hide, Graham argued, is their text: the characteristic language of spam. They can forge return addresses and headers and send their messages through proxies and open relays and so on, but that distinctive spammy tone, the cajolery or the plea, must be there to convince the human on the far side to click the link. What he proposed was a method for turning that language into an object available to hackers, making it possible to build near-perfect personal filters that can rapidly improve in relation to spam with the help of mathematical tools.

The Bayesian probability that Graham adopted for this purpose is named for Thomas Bayes, who sketched it out in the 1750s. It can be briefly summarized by a common analogy using black and white marbles. Imagine someone new to this world seeing the first sunset of her life. Her question: will the sun rise again tomorrow? In ignorance, she defaults to a fifty-fifty chance and puts a black marble and a white marble into a bag. When the sun rises, she puts another white marble in. The probability of randomly picking white from the bag—that is, the probability of the sun rising based on her present evidence—has gone from 1 in 2 to 2 in 3. The next day, when the sun rises, she adds another marble, moving it to 3 in 4, and so on. Over time, she will approach (but never reach) certainty that the sun will rise. If, one terrible morning, the sun does not rise, she will put in a black marble and the probability will decline in proportion to the history of her observations. This system can be extended to very complex problems in which each of the marbles in the bag is itself a bag of marbles: a total probability made up of many individual, varying probabilities—which is where email, and spam, come into the picture.

A document, for a Bayesian filing program, is a bag—a total probability—containing many little bags. The little bags are the tokens—the words—and each starts out with one white marble and one black. You train the program by showing it that this document goes in the Mail folder, and that document in the Spam folder, and so on. As you do this with many documents, the Bayesian system creates a probability for each significant word, dropping in the marbles. If it makes a mistake, whether a "false positive" (a legitimate message marked as spam) or a "false negative" (spam marked as a legitimate message), you correct it, and the program, faced with this new evidence from your correction, slightly reweights the probabilities of the words found in that document for the future. It turns language into probabilities, creating both a characteristic vocabulary of your correspondence and of the spam you receive. It will notice that words like "madam," "guarantee," "sexy," and "republic" almost never appear in legitimate mail and words like "though," "tonight," and "apparently" almost never appear in spam. Soon it will be able to intercept spam, bouncing it before it reaches your computer or sending it to the spam folder. The Bayesian filer becomes the Bayesian filter.[19]

This idea was hardly new when Graham posted his essay. For somewhat mysterious reasons, naïve Bayesian algorithms happen to be exceptionally good at inductive learning tasks such as classifying documents and had been studied and applied to mail before.[20] Jason Rennie's ifile program—the one for which he'd volunteered his own email for testing purposes—was applying naïve Bayes to filing email and discarding "junk mail" in 1996. "As e-mail use has grown," he writes his in 1998 paper on ifile, "some regularity has come about the sort of e-mail that appears in users' mail boxes. In particular, unsolicited e-mail, such as 'make money fast' schemes, chain letters and porn advertisements, is becoming all too common. Filtering out such unwanted trash is known as junk mail filtering."[21] That same year, five years before Graham's essay, several applications of Bayesian filtering systems to email spam were published; the idea was clearly in the air. Why didn't it take off, then? The answer to that question explains both what made Graham's approach both so successful and where its weakness lay in ultimately stopping spam.

Understanding that answer requires a dive into some of the technical material that underlies the project of filtering. The biggest problem a filter can have is called *differential loss*: filtering the "false positives" mentioned earlier, when it misidentifies real mail as spam and deletes it accordingly. Email deals with time-dependent and value-laden materials such as job offers, appointments, and personal messages needing prompt response, and the importance of messages varies wildly. The potential loss can be quite different from message to message. (Think of the difference between yet another automatic mailing list digest versus a letter from a long-lost friend or work from a client.) The possibility of legitimate email misclassified as spam and either discarded or lost to the human eye amid hundreds of actual spam messages is so appalling that it can threaten to scuttle the whole project. In fact, setting aside the reality of potential losses in money, time, and miscommunication, the psychological stress of communicating on an uncertain channel is onerous. Did he get my message, or was it filtered? Is she just ignoring me? Am I missing the question that would change my life? It puts the user in the classic epistemological bind—you don't know what you're missing, but you know that you don't know—with the constant threat of lost invitations, requests, and offers. With a sufficiently high rate of false positives, email becomes a completely untenable medium constantly haunted by failures of communication.

The loss in time paid by someone manually refiling a spam message misclassified as legitimate and delivered is quite small—about four seconds, on average—especially when compared to the potential disaster and distress of false positives.[22] Filters, therefore, all err a little to the side of tolerance. If you set the filter to be too restrictive, letting through only messages with a very high probability of being legitimate, you run the risk of unacceptably high numbers of false positives. You have to accept messages on the borderline: messages about which the filter is dubious. This is the first strike against the early Bayesian spam filters, part of the answer to Graham's question, "If people had been onto Bayesian filtering four years ago, why wasn't everyone using it?"[23] Pantel and Lin's Bayesian system SpamCop had a rate of 1.16 percent false positives to Graham's personal rate of 0.03 percent. Small as it may seem, spammers could take advantage of that space and the anxiety it produced.

Perhaps you can narrow that rate a bit by adding in some known spam characteristics—reliable markers that will separate spam from your mail. One group of early spam filter designers tried this method, and the collection of properties built into their filter to specify spam are a fascinating artifact, because the intervening decade has rendered them almost entirely wrong.[24] They made an assumption of stability that spammers turned into a weakness, an issue that Graham's ever-evolving system sidesteps. The programmers chose two kinds of "domain specific properties" or things that marked spam as spam: thirty-five phrases ("FREE!," "be over 21," "only $" as in "only $21.99") and twenty nonphrasal features. The language of spam has obviously changed enormously—not least as a result of Bayesian filtering, as we will see—but the twenty additional features are where its mutability comes across the most. They include the percentage of non-alphanumeric characters (such as $ and !), "*attached* documents (most junk E-mail does not have them)," "when a given message was received (most junk E-mail is sent at night)," and the domain type of the sender's email address, as "junk mail is virtually never sent from .edu domains."

The "no attachments" rule does not hold: many varieties of spam include attachments, including viruses or other malware, in the form of documents and programs (as well as more baroque attachments such as mp3 files purporting to be voicemail that act as come-ons for identity theft exploits). Neither does reliance on delivery times, which have become much more complex, with interesting diurnal cycles related to computers being turned on and off as the Earth rotates. Address spoofing (by which messages can appear to come from any given address) and commandeered academic addresses (which are usually running on university servers with high-bandwidth connections, great for moving a few million messages fast) has rendered .edu addresses fairly meaningless. (Even worse was an oversight in the SpamAssassin filter that gave any message sent in 2010 a very low score—flagging it as likely to be spam—because that year "is grossly in the future" or was, when the filter was being developed.)[25]

These fixed filtering elements, rendered not just useless but misleading after only a few years, highlight perhaps the biggest hurdle for the scientific antispam project: it moved so slowly in relation to its quarry. Formally accredited scientists publish and are citable, with bracketed numbers directing us to references in journals; hackers, self-described, just post their work online, however half-baked or buggy, because other people will help fix it. Compared to the patient corpora-building and discussion of naïve Bayesian variants in the scientific antispam project, the hackers' antispam initiative was almost hilariously ramshackle, cheap, fast, semifunctional, and out of control. Graham's "plan for spam" was written before he came across extant research papers, but it addresses the problems they faced. (In a later talk, he brought up his algorithm for lazy evaluation of research papers: "Just write whatever you want and don't cite any previous work, and indignant readers will send you references to all the papers you should have cited.")[26] It offers a much lower rate of potential false positives and an entirely individual approach to training the system to identify spam—one that eliminates the need for general specifications. In theory, it acts as a pure, automated reflection of spam, moving just as fast as the ingenuity of all the spammers in the world. As they try new messages, it learns about them and blocks them, and making money at spam gets harder and harder. Above all, Graham's approach was about speed.

His essay reflects this speed. There are no equations, except as expressed in the programming language Lisp—the question is not "Does this advance the mathematical conversation?" but "Does it run, and does that make a difference?" There are no citations, though he thanks a few people at the end (the "lazy evaluation of research papers" method kicked in afterward, when his essay had been mentioned on the geek news site Slashdot). The language of process and experiment is completely different: "(There is probably room for improvement here.) . . . [B]y trial and error I've found that a good way to do it is to double all the numbers in *good*. . . . There may be room for tuning here . . . I've found, again by trial and error, that .4 is a good number to use." It offers a deeply different intellectual challenge than the papers described previously, saying, in essence, "This is what I did. If you think you can improve it, open a terminal window and get to work." It is a document published in a wizardly environment in which anyone, at least in theory, can be a peer with the right attitude and relevant technical skills—and where the review happens after the fact, in linkback, commentary and further development of the project.

It was received with this same sense of urgency and hands-on involvement, starting an avalanche of critique, coding, collaboration, and commentary. One example among many took place on the Python-Dev mailing list discussion (Python is a high-level programming language, named, as it happens, for *Monty Python*). They were discussing bogofilter, a spam filter that applied Graham's naïve Bayes model: "Anybody up for pooling corpi (corpora?)?" writes a programmer almost immediately after Graham posted his essay. Tim Peters brings mathematical clarity to the table: "Graham pulled his formulas out of thin air, and one part of the scoring setup is quite dubious. This requires detail to understand"—which he then provides, moving through a deep problem in the application of Bayes's theorem in Graham's model.[27] Open source advocate Eric Raymond replies with a possible workaround, closing his message with a question for Peters: "Oh, and do you mind if I use your algebra as part of bogofilter's documentation?" Peters replies: "Not at all." Within days the bogofilter project, based on Graham's naïve Bayesian idea, was checked into a testing environment for developing software. The pace of spam's transformations seemed to have found its match.

To really understand the impact of the naïve Bayesian model on email and on spam, multiply the Python bogofilter project's fast-paced collaborative, communal effort out into several different programming languages and many competing projects. Graham lists a few in the FAQ he wrote following "A Plan for Spam": Death2Spam, SpamProbe, Spammunition, Spam Bully, InboxShield, Junk-Out, Outclass, Disruptor OL, SpamTiger, JunkChief. Others come up in a separate essay: "There are now over 30 available. Apple has one, MSN has one, AOL is said to have one in beta, and you can be pretty sure Yahoo is working on one."[28] The Graham-promoted model of naïve Bayesian filtering remains the default approach to antispam filters to this day, though of course with many modifications, additions, and tweaks, as the Python-Dev conversation suggested. Very heavily modified and supplemented naïve Bayesian filters operate both at the level of personal email programs and the level of the very large webmail providers such as Microsoft's Hotmail and Google's Gmail (in concert with numerous other filtering techniques).

There are many reasons for its success. There's context, for one: Graham's was a chatty essay full of code examples posted by a widely read Internet entrepreneur, then linked and reposted on high-traffic news and discussion sites, rather than a technical research paper presented at an academic workshop about machine learning for text classification. With its language of "probably room for improvement here," it was built for rapid adoption by influential groups in the hacker community, like the gang at Python-Dev. Most important, it offered a thoughtful and persuasive argument for its graceful technical hack, an argument about the social and economic structure of spam—how spam works in the most general sense and how it could be broken. This argument was adopted, implicitly and explicitly, with Graham's filter, and, along with the problem of false positives, it formed the seam of failure that spam practices exploited in their transformation and survival.

All of Graham's argument in the original "A Plan for Spam" hinges on two points. The first is that "the Achilles heel of the spammers is their message." The only point on which to reliably stop spam—the one thing spammers cannot circumvent—are the words they need to make the recipient act in some way. Hence the use of Bayesian filters to analyze the text itself and to treat the very language used in the appeal to block it. The second, related point is that this block does not need to work perfectly but just very well, because the goal is not to criminalize spam, publicly shame spammers, or educate recipients, as previous projects sought to—the goal is simply to make spamming less profitable. "The spammers are businessmen," he writes. "They send spam because it works. It works because although the response rate is abominably low . . . the cost, to them, is practically nothing. . . . Sending spam does cost the spammer something, though. So the lower we can get the response rate—whether by filtering, or by using filters to force spammers to dilute their pitches—the fewer businesses will find it worth their while to send spam." The promise is that spam is that rare class of problem that, when ignored, actually goes away. "If we get good enough at filtering out spam, it will stop working, and the spammers will actually stop sending it."[29]

Graham saw legitimated and theoretically law-abiding attempts at email marketing as spam's most dangerous wing, the leading edge of an Internet in which spammers who can afford lobbyists work with impunity and email has become one big marketing machine. Opt-in spam was one of the most popular forms of this going-legit spam activity, whose proponents argued that the recipients of their spam messages had subscribed—usually by having entered their address at a website whose terms of service allowed the site's owners to sell addresses to marketers. Opt-in spam included courtesy instructions, at least in theory, about how to unsubscribe from the mailings. "Opt-in spammers are the ones at the more legitimate end of the spam spectrum," Graham wrote. "The arrival of better filters is going to put an end to the fiction of opt-in, because opt-in spam is especially vulnerable to filters [due to the readily recognizable legal and unsubscribing boilerplate text that appeared in the messages]. . . . Once statistical filters are widely deployed, most opt-in spam will go right into the trash. This should flush the opt-in spammers out of their present cover of semi-legitimacy."[30] If people could vote with their filters as to what was spam, you would have to start actively deceiving them and engaging in less legal and more lucrative practices to make it pay. With such filters broadly deployed, "online marketers" striving for legitimacy could no longer take shelter under the vague protection of legal precedents for telemarketing and direct mail. They would either drop out of the business or become entirely criminal, with all the additional personal, social, and financial costs that change entails. You did not need to criminalize them, because you could make them criminalize themselves—and then you could sic the law on them.

"The companies at the more legitimate end of the spectrum lobby for loopholes that allow bottom-feeders to slip through too. . . . If the 'opt-in' spammers went away . . . [i]t would be clear to everyone where marketing ended and crime began, and there would be no lobbyists working to blur the distinction."[31] Graham had thus fashioned an extraordinary double bind. The legal specifications that could legitimate spam and the materials used to display a message's compliance with a law such as CAN-SPAM—disclaimers, citations of the relevant laws, statements of compliance, and links to unsubscribe—are very regular and therefore a perfect target for Bayesian filtering. With good filters in place, the process of becoming legitimate—of complying with the laws—makes your business much less profitable. You have to become a criminal or get out.

Along with high-profile busts and legal cases in the American spam community such as those against Sanford Wallace and Alan Ralsky, Graham's filter and its argument succeeded almost completely. Laws and filtering together wiped out the world of legitimated online marketers, the kind of people who wanted to move lots of relatively low-margin products with a straightforward pitch to the world's email addresses, by starving them in a deep financial freeze. In the process the field was left to agile, adaptable, and far more resourceful criminals. In Graham's plan for spam lay the seeds of its own relative undoing. In its success, it killed the profit model most vulnerable to it and initiated the transformation of spam into a new and still more problematic practice, one far harder to control.

POISONING: THE REINVENTION OF SPAM

INVENTING LITSPAM

The machines in the shop roar so wildly that
often I forget in the roar that I am; I am
lost in the terrible tumult, my ego disappears, I
am a machine. I work, and work, and work without end; I am busy, and busy, and busy at all time.
For what? and for whom? I know not, I ask not!
How should a machine ever come to think?

—Morris Rosenfeld, "In the Sweat-Shop" (*Songs from the Ghetto*, trans. Leo Wiener, Norbert Wiener's father)

Even as the filters were being installed, the first messages began to trickle in, like this one: "Took her last look of the phantom in the glass, only the year before, however, there had been stands a china rose go buy a box of betel, dearest of the larger carnivorous dinosaurs would meet it will be handy? Does it feel better now?" Or like this: "rosily pixie sir.chalet, healer partly.fanned media viva.jests, wheat skier.given rammed bath.weeded divas boxers." Messages by the hundreds and thousands, sometimes with a link, mostly without. It was as though some huge Dada machine, the Tzara-Bot 9000, had just come online. This was *litspam*, cut-up literary texts statistically reassembled to take advantage of flaws in the design and deployment of Bayesian filters.

Bayesian filters destroyed email spam as a reputable business model in three ways, each of which became a springboard for spam's transformation. Filtering killed off conventional spam language, the genre of the respectable sales pitch, with its textual architecture of come-on rhetoric inherited from generations of Dale Carnegie books, direct-mail letters, cold calls, and carnival barkers ("Hundreds of millions of people surf the Web today. The internet is absolutely exploding everywhere in the world. Ask yourself: 'Am I going to profit from this?' Give me the chance to share with you this exiting business opportunity."). This kind of material became a liability; its elements were too likely to be caught by the statistical analysis of words that the filters performed. Second, filtering made it a lot harder to make money through sales—if far fewer of your messages were getting through, you needed much better return on a successful message, not just a small profit on a bottle of generic pharmaceuticals, to make spam a viable business. Finally, filtering enormously increased the failure rate for spammy messages. If the filter caught the vast majority of your messages, you needed to send out a lot more, and be much more creative in their construction, to turn that tiny percentage or part of a percentage into a business. It was assumed that the message-sending capacity was a reliable limit, a fixed ceiling on spam operations: in the "Plan for Spam FAQ," Graham answers the question "If filters catch most spam, won't spammers just send more to compensate?" with "Spammers already operate at capacity." These three developments fed each other. If the filters attacked regularity in language, noting the presence of words with a high probability of appearing in spam messages, you had to be much more creative in the spam messages you wrote, putting more effort into each attempt than before. You would see very little profit in return for your increased effort, because fewer of your messages would get through, and you would have to put more of that profit into your infrastructure, because you would need to hugely increase the amount of spam you could send.

They also contained three points for spam's transformation into a new and different trade, of which litspam was a harbinger. All three points of transformation hinged on the success of Graham's ideas: the enforcement of new laws combined with filtering had eliminated the mere profit-seeking carpetbaggers and left the business to the criminals. Filters made conventional sales language and legal disclaimers into liabilities, which meant that those willing to be entirely duplicitous could move into wholly different genres of message to get past the filters and make the user take action. Hence the recommended link from a half-remembered friend (or friendly stranger), the announcement of breaking news, and, most extraordinarily, the fractured textual experiment of litspam. If filtering made it much harder to make money per number of messages, spam messages could become much more individually lucrative: rather than sales pitches for goods or sites, they could be used for phishing, identity theft, credit card scams, and infecting the recipient's computer with viruses, worms, adware, and other forms of dangerous and crooked malware. A successful spam message could net many thousands of dollars, rather than $5 or $10 plus whatever the spammer might make selling off their good addresses to other spammers. Finally, if the new filters meant the messages failed much more often, spammers could develop radically new methods for sending out spam that cost much less and enabled them to send much, much more—methods such as botnets, which we'll come to shortly.

In the context of the transformations that spam was going through as it became much more criminal, experimental, and massively automated, litspam provides a striking example of the move into a new kind of computationally inventive spam production. Somewhere, an algorithmic bot with a pile of text files and a mailing list made a Joycean gesture announcing spam's modernism.

To explain litspam, recall the problem of false positives: legitimate messages misfiled as spam. You cannot make the filter too strict. You need to give it some statistical latitude, because missing a legitimate message could cost the user far more than the average of 4.4 seconds it takes to identify and discard a spam message that gets through the filter. The success or failure of a filter depends on its rate of false positives; one important message lost may be too many, and Graham argued that the reason Bayesian filters did not take off in their first appearance was false positive rates like Patel and Lin's 1.16 percent, rather than his 0.03 percent. Implicit in his argument was the promise that other people could reproduce or at least closely approximate his percentage. A person could indeed reproduce Graham's near-perfect rate of false positives if they were very diligent, particularly in checking the marked-as-spam folder early in the filter's life to correct misclassifications. It also helped to receive a lot of email with a particular vocabulary, a notable lexical profile, to act as the "negative," legitimate nonspam words. These things were true of Graham. Building this filter was a serious project of his for which he was willing to read a lot of spam messages, do quite a bit of programming, and become a public advocate; it follows that his personal filter would be very carefully maintained. Graham had a distinctive corpus to work with on the initial filter: his personal messages, with all the vocabulary specific to a programmer and specialized venture capitalist—"Lisp [the programming language] . . . is effectively a kind of password for sending email to me," he wrote in the original "A Plan for Spam" document. His array of legitimate words, at the opposite side of the axis from the words that signal spam (such as "madam," "guarantee," and "republic"), includes words like "perl [another programming language]," "scripting," "morris," "quite," and "continuation."

Other individual users, however, may have a slightly higher rate of false positives because they have a characteristic vocabulary that overlaps more with the vocabulary of spam than Graham's does, or because their vocabulary is aggregated with that of others behind a single larger filter belonging to an organization or institution, or simply because they are lazier in their classifications or unaware that they can classify spam rather than deleting it. (The problem of characteristic vocabulary was even worse for blog comment spam messages—the kind with a link to boost Google search rankings or bring in a few customers—because the spammers, or their automated programs, can copy and cut up the words in the post itself for the spam massage, making evaluation much trickier.) Users are thus not perfect, and filters can be poorly implemented and maintained, and so must be a little tolerant of the borderline messages. In this imprecision, a two-pointed strategy for email spamming took shape:

1. In theory, you could sway a filter by including a lot of neutral or acceptable words in a message along with the spammier language, edging the probability for the message over into legitimacy. Linkless gibberish messages were the test probes for this idea, sent out in countless variations to see what was bounced and what got through: "I although / actually gorged with food, were always trumpets of the war / to sound! This stillness doth dart with snake venom on itwell, / I'd have laughed."
2. After a spam message got through, the recipient was faced with a dilemma. If the recipient deleted the message, rather than flagging it as spam, the filter would read it as legitimate, and similar messages would get through in the future. If he or she flagged it as spam, the filter, always learning, would add some more marbles to the bags of probabilities represented by significant words, slightly reweighing innocent words such as "stillness," "wheat," "laughed," and so on toward the probability of spam, cumulatively increasing the likelihood of false positives. These broadcasts from Borges's Library of Babel would be, in effect, a way of taking words hostage. "Either the spam continues to move, or say goodbye to 'laughed.'"

But why use literature? Early messages show that the first experiments along these lines were built of words randomly drawn from the dictionary. This approach does not work very well, because we actually use most words very seldom. The most frequently used word in English, "the," occurs twice as often as the second most frequent, and three times as often as the third most frequent, and so on, with the great bulk of language falling far down the curve in a very long tail.[32] From the perspective of the filter, all those words farther out on the curve of language—"abjure," "chimera," "folly"—are like the bag of marbles after that first sunset, with one black marble and one white; with no prior evidence, those unused words are at fifty/fifty odds and make no difference, and one "sexy" will still flag the message as spam. What the spammer needs is natural language, alive and in use up at the front of the curve.

A huge portion of the literature in the public domain is available online as plain text files, the format most convenient to programmers: thousands and thousands and thousands of books and stories and poems. These can be algorithmically fed into the maw of a program, chopped up and reassembled and dumped into spam messages to tip the needle just over into the negative, nonspam category. Hence the bizarre stop-start rhythm of many litspam messages, with flashes of lucidity in the midst of a fugue state, like disparate strips of film haphazardly spliced together. Their sources include all the canonical texts and work in the public domain available on sites like Project Gutenberg, as well as more recondite materials. Many authors in the science fiction vein are popular with hackers, who sometimes pay them the dubious honor of scanning their books with optical character recognition software to turn the printed words into text files that can be circulated online. Neal Stephenson's encryption thriller *Cryptonomicon* is one of these books, available as a full text file through a variety of sources and intermittently in the form of chunky excerpts in spam messages over the course of years. "This is a curious form of literary immortality," Stephenson observed. "E-mail messages are preserved, haphazardly but potentially forever, and so in theory some person in the distant future could reconstruct the novel by gathering together all of these spam messages and stitching them back together. On the other hand, e-mail filters learn from their mistakes. When the *Cryptonomicon* spam was sent out, it must have generated an immune response in the world's spam filtering systems, inoculating them against my literary style. So this could actually cause my writing to disappear from the Internet."[33]

The deep strangeness of litspam is best illustrated by breaking a piece of it down, dissecting one of these flowers of mechanized language. The sample that opened this section, drawn at random from one of my spam-collecting addresses is two sentences and forty-five words and was assembled from no less than four interpolated sources: "Took her last look of the phantom in the glass, only the year before, however, there had been stands a china rose go buy a box of betel, dearest of the larger carnivorous dinosaurs would meet it will be handy? Does it feel better now?" "Took her last look of the phantom in the glass" is from "The Shadows," a fairy tale by Aberdonian fantasist George MacDonald. "Only the year before, however," and "of the larger carnivorous dinosaurs would meet" are from chapters 15 and 11 of Arthur Conan Doyle's adventure novel *The Lost World*. "Stands a china rose go buy a box of betel, dearest" is from Song IV of the "Epic of Bidasari" as translated in the Orientalist Chauncey Starkweather's *Malayan Literature*. And "it will be handy? Does it feel better now?" is from Sinclair Lewis's *Main Street*, chapter 20. Each of these fragments are subtly distorted in different ways—punctuation is dropped and the casing of letters changed—but left otherwise unedited. It's a completely disinterested dispatch from an automated avant-garde that spammers and their recipients built largely by accident. "I began to learn, gentlemen," as the ape says in Kafka's "Report to an Academy," another awkward speaker learning language as a means of escape: "Oh yes, one learns when one has to; one learns if one wants a way out; one learns relentlessly."[34]

Litspam obviously does not work for human readers, aside from its occasional interesting resemblance to stochastic knockoffs of the work of Tzara or Burroughs (with a hint of Louis Zukofsky's quotation poems, or Bern Porter's "Founds" assembled from NASA rocket documentation). If anything, its fractured lines and phrasal salad are a sign that something's suspiciously wrong and the message should be discarded. As with the biface, robot-readable text of web pages that tell search engine spiders one thing and human visitors another, litspam is to be read differently by different actors: the humans, with their languages, and the filters with their probabilities, like the flower whose color and fragrance we enjoy, and whose splotched ultraviolet target the bee homes in on. Litspam cuts to the heart of spam's strange expertise. It delivers its words at the point where our experience of words, the Gricean implicature that the things said are connected in some way to other things said or to the situation at hand, bruisingly intersects the affordances of digital text. Like a negative version of the Turing test, you think you will be chatting with a person over teletype ("Will X please tell me the length of his or her hair?" as Turing suggests) and instead end up with racks of vacuum tubes or, rather, a Java program with most of English-language literature in memory: "that when some members rouen, folio 1667.anglonorman antiquities p. Had concluded his speech to the king."[35] We look for sense, for pattern and meaning, whether in the Kuleshov Effect—the essence of montage, with different meanings attributed to the same strip of film depending on what it's intercut with—or the power of prophetic signals, like a spread of Tarot cards, whose rich symbolic content is full of hooks we can connect with our own current preoccupations, fears, memories, and desires. If there's a spammy core to the message—a recognizable pitch, link, or come-on—we might pick out that most salient portion (perhaps clicking on this will explain this bizarre message!) and the spam will still have done its job.

Let us return to Turing, briefly, and introduce the fascinating Imitation Game, before we leave litspam and the world of robot-read/writable text. The idea of a quantifiable, machine-mediated method of describing qualities of human affect recurs in the literature of a variety of fields, including criminology, psychology, artificial intelligence, and computer science. Its applications often provide insight into the criteria by which different human states are determined—as described, for example, in Ken Alder's fascinating work on polygraphs, or in the still understudied history of the "fruit machine," a device that (allegedly) measured pupillary, pulse, and other response to pornographic images, developed and deployed during the 1950s for the purpose of identifying homosexuals in the Canadian military and the Royal Canadian Mounted Police (RCMP) in order to eliminate them from service.[36] (It is like a sexually normative nightmare version of the replicant-catching Voight-Kampff machine in *Blade Runner.*) Within this search for human criteria, the most famous statement—and certainly the one that has generated the most consequent literature—is the so-called Turing Test. The goal of Turing's 1950 thought experiment (which bears repeating, as it's widely misunderstood today) was to "replace the question [of 'Can machines think?'] by another, which is closely related to it and is expressed in relatively unambiguous words."[37] Turing considered the question of machines "thinking" or not to be "too meaningless to deserve discussion," and, quite brilliantly, turned the question around to whether *people* think—or rather how we can be convinced that other people think. This project took the form of a parlor game: A and B, a man and a woman, communicate with an "interrogator," C, by some intermediary such as a messenger or a teleprinter. C knows the two only as "X" and "Y"; after communicating with them, C is to render a verdict as to which is male and which female. A is tasked with convincing C that he, A, is female and B is male; B's task is the same.[38] "We now ask the question," Turing continues, "'What will happen when a machine takes the part of A in this game?' Will the interrogator decide wrongly as often when the game is played like this as he does when the game is played between a man and a woman? These questions replace our original, 'Can machines think?'"

What litspam has produced, remarkably, is a kind of parodic imitation game in which one set of algorithms is constantly trying to convince the other of their acceptable degree of salience—of being of interest and value to the humans. As Charles Stross puts it, "We have one faction that is attempting to write software that can generate messages that can pass a Turing test, and another faction that is attempting to write software that can administer an ad hoc Turing test."[39] In other words, what we are seeing is the product of algorithmic writers producing text for algorithmic readers to parse and block, with the end product providing a fascinatingly fractured and inorganic kind of discourse, far off even from the combinatorial literature of avant-garde movements such as the Oulipo, the "Workshop of Potential Literature." The particular economics of spamming reward sheer volume rather than message quality, and the great technical innovations lie on the production side, building systems that act with the profligacy of a redwood, which may produce a billion seeds over the course of its lifetime, of which one may grow into a mature tree.[40] The messages don't improve from their lower bound unless they have to, so the result doesn't get "better" from a human perspective—that is, more convincing or plausibly human—just stranger.

Surrealist automatic writing has its particular associative rhythm, and the Burroughsian Cut-Up depends strongly on the taste for jarring juxtapositions favored by its authors (an article from *Life*, a sequence from *The Waste Land*, one of Burroughs's "routines" in which mandrills from Venus kill Eisenhower). Litspam text, along with early comment spam and the strange spam blogs described in the next section, is the expression of an entirely different intentionality without the connotative structure produced by a human writer. The results returned by a probabilistically manipulated search engine, or the poisoned Bayesian spew of bot-generated spam, or the jumble of links given by a bad choice filtering algorithm act in a different way than any montage. They are more akin to flipping through channels on a television, with very sharp semantic transitions between spaces—from poetry to porn, a wiki to a closed corporate page, a reputable site to a spear-phishing mock-up. (If it has a cultural parallel, aside from

John Cage's *Imaginary Landscape No. 4*—in which two musicians manipulate a radio's frequency, amplitude, and timbre according to a preestablished score, with no control over what's being broadcast—it would be Stanley Kubrick's speculative art form of the future, which he described as "mode jerking": sudden, severe, jolting transitions between different environments.)[41] Consider this message from "AKfour seven," writing via a Brazilian domain hosted on an ISP in Scranton, Pennsylvania (ignore the HTML markup):

> <p>I stand here today humbled by the task before [url=http://www.bawwgt.com]dofus kamas[/url], grateful for the trust you have bestowed, mindful of the sacrifices borne by our [url=http://www.bawwgt.com]cheap dofus kamas[/url]. I thank President [url=http://www.bawwgt.com]dofus power leveling[/url] for his service to [url=http://www.bawwgt.com]buy dofus kamas[/url], as well as the generosity and cooperation he has shown throughout this transition.</p>

It's President Obama's inaugural address, intercut with links to a site whose business it is to sell currency ("Kamas") and other desirables for the French online role-playing game Dofus, which features various tree people, archers, and gambling cats—and a substantial gray market for in-game currency sold for real money. It is not simply that the smallest is spliced into the biggest, major with minor, now with then. It is the use of written words under the condition of pure arbitrary utility. As digitized, searchable, copy-and-paste-ready text, it is all one continuous matter—almost shockingly atemporal and best analogically compared not to a library or a conversation but to the "polymer goo" that Harrison White uses to describe social structures, full of complex striations and from which many different shapes can be extruded depending on need.[42]

Finally, a note on this idea of "atemporality" to close this section on litspam. The concept of atemporal media has been discussed recently in terms of digital aesthetics and music. The digitization of media moves them into a kind of continuous present of use, the way all recorded music can now occupy a single, shuffled state of immanence from wildly different points of creation. An mp3 of the antediluvian old-time musician Dock Boggs, as recorded in the Norton Hotel with a borrowed banjo in 1927, segues into the synthesizer layers of Oneohtrix Point Never, who creates music in 2010 that could pass for electrocosmic voyages on vinyl from the 1970s. Historicity becomes another stylistic element, like timbre. As Brian Eno has put it, it's all "current" now, which brings the aesthetics of recording itself to the fore as a stylistic choice with its own content, as all sounds coexist in a permanent digital noon. Litspam, setting aside its ultimate purpose of slipping through or damaging filters to sell more porn site logins or discontinued toys, is an extraordinary form of digital atemporality. Histories and myths, poetry, instructions for pleaching the lime trees of an ornamental garden, religious exegesis, and online tax guides constitute one shape, of which a given litspam message is a probability-guided surface. "In which gravitation is a consequence of the curvature of spacetime which governs the motion of inertial objects. The South Park episode Conjoined Fetus Lady and Season 1 of Freaks and Geeks depict dodgeball as a potentially violent sport. August Anheuser Busch IV (born June 15, 1964) is the great-great-grandson of Anheuser-Busch founder Adolphus Busch, the son of former chairman, president and CEO August Busch III. Many of these are produced by hurricanes or tropical storms along the coastal plain."

Litspam was only one new form of postfilter spam among many, of course. Graham predicted that "the spam of the future will probably look something like this: 'Hey there. Thought you should check out the following: http://www.27meg.com/foo.'" It squeaks by the filter on neutral language but gets caught with its suspicious URL, and we have indeed seen quite a bit of that, along with the variety of old-fashioned spam that makes it past imperfectly installed and trained filters.[43] (Filtering also created a genius for euphemism on the part of spammers. A few of many, many terms in recent spam messages for promises about the male anatomy, which almost approach poetic allusion: "your engine in pants"/"Drilling machine"/"'in-work-condition' tool"/"Crazy penetrator!"/"Meatstick-champion!"/"Your nighttime failure"/"Make your volcano erupt over lion"/"the thing as you deserve it" and many others). Litspam remains something remarkable and special as an unintended consequence within the unintended consequence of spam: a loop of mechanical readers and mechanical writers generating texts from within the uncanny valley identified by Masahiro Mori. It is the chance meeting of *Ulysses* and a telephone interchange, as strange to our eyes as a pedantic speech from an ape, a tale told by a robot.

THE NEW SUCKERS

Graham never claimed that he or anyone else could filter spam perfectly, only that the filters would work well enough to make spamming into an unprofitable business. The various flavors of Bayesian filtering did, in fact, massively curtail the delivery of spam to the world's inboxes. ISPs have the first layers of filters between the individual mailbox and the rest of the network, and by end of 2006, they saw spam become an estimated 85 percent of all mail traffic on the far side of their filters—a number that holds steady, give or take a few percentage points, to this day.[44] Most people see only a minuscule portion of this total amount. A vast wave is crashing continuously against the filters, with some occasional spillover. This was exactly the plan that Graham outlined. The response rate for spam was always dismal: spammer Davis Hawke reported a decent return at two-tenths of 1 percent in the period prior to the widespread use of Bayesian filters, and those filters hugely cut down on the amount delivered.[45] It worked, then, on its technical terms and only on those terms. Therein lay spam's vector. The social choices embedded in and enabled by the technology became the points of failure. In retrospect, these critical points were four: two on the side of users and two on the side of spammers.

FILTERS WERE UNEVENLY DEPLOYED AND TRAINED

Some ISPs, organizations, and users will do it better than others; some possess a more distinctive vocabulary; and some are more diligent in managing the training of the system. There will be varying estimations of the cost and acceptable chance of false positives. Many users may never be aware of the need to "classify-as-spam" when sorting their incoming mail. Rates will vary, programs will become obsolete, and there will be holes, however small.

THE QUESTION OF THE "15 IDIOTS"

Graham, in the months following "A Plan for Spam," considered the possibility that the people most susceptible to spam—the people that make it profitable—will overlap with those least likely to install filters or feel comfortable using them. Arguing that spam makes money from the "15 stupidest or most perverted" people in a million, Graham continues: "The great danger is that whatever filter is most widely deployed in the idiot market will require too much effort by the user. . . . [T]he 15 idiots are probably also the 15 users who won't bother."[46] His solution, such as it is, is an assumption in the "Plan for Spam FAQ": "I suspect that people stupid enough to respond to spam will often get email through one of the free services like Yahoo Mail or Hotmail, or through big providers like AOL or Earthlink. Once word spreads that it is possible to filter out most spam, they'll be forced to offer effective filters."[47]

CHANGE IN THE ECONOMICS OF SPAM PRODUCTION AND DISTRIBUTION

"Spammers," Graham averred, "already operate at capacity." In fact, as the filters went online, the production of the spam they were trying to stop was changing. The end of the legitimated spammers in a double bind of filtering and changes to the law was one rumble in spam's tectonic shift toward an almost wholly criminal domain. Abandoning any pretense of legitimacy freed up a great deal of technical ingenuity. The development of systems like botnets, the use of ISPs in foreign jurisdictions (in some cases owned outright by gangsters), and the increasing sophistication of the programming of spam software at once boosted the capacity of spam distribution while lowering operating costs.

LIBERATION OF SPAM INTO PURE EXPERIMENT—AND PURE FRAUD

In changing its business model, the criminalization of spam also changed its arsenal of tools and words. It no longer needed even the pretense of selling products in a way that would appear legally reputable. Strategies such as phishing and identity theft, advance-fee fraud, and virus and malware distribution meant that the profit margin was pushed back up as the cost of distribution dropped, and spam started to sound, linguistically, like many things—some of them never heard before. Free to abandon any trappings of genre, it could seek any textual shape that got past the filters, using Shakespeare the way a bacterium consumes and employs foreign DNA, making spam into a different and a weirder beast than the one Bayesian filters had been designed to stop.

These problems are related. When Graham described the "15 idiots" as "stupid" or "perverted," he was writing, with hacker hauteur, about people responding to messages that seemed to require enormous gullibility or a great fondness for porn. But the move into full-on criminality increased the pool of potential suckers. Many people who would never respond to an ad for a multilevel marketing scheme would respond to a notice purporting to be from their bank or PayPal account. Spam could now more aggressively target the old, the confused, people using the Internet in a second language, and new users in general, putting the sting on the *naïfs* that naïve Bayes was supposed to protect. You no longer needed to be an idiot to be one of the fifteen idiots, and this meant that each new sucker could be worth a lot more than the old. This money in turn attracted more sophisticated and skilled talent to spam, both on the business and the programming sides—the kind of people who could build more complex litspam engines and spam distribution programs. Graham, and those who preceded and followed him in the search for a probabilistic filter, were building a brilliant hack to solve a complex and embedded problem whose elements were simultaneously technical and social at every step. The social element transformed in response to their technical intervention and altered the technical in turn.

This is only the merest introduction to the army of chattering, crude language machines ("It was so easy to imitate these people," as Kafka's ape says: "I could already spit on the first day") produced by spammers. To take the measure and extent of their population, we have to turn to the world of spam blogs geared to influence search engines—built to beat a whole different order of filters, the avant-garde in the art of misdirection.

"NEW TWIST IN AFFECT": SPLOGGING, CONTENT FARMS, AND SOCIAL SPAM

THE POPULAR VOTE

Screams of frightened women, choked Sobs, truly communicative Tears, little brusque Laughs . . . Howls, Chokings, Encore!, Recalls, silent Tears, Threats, Recalls with additional Howls, Pounding of approbation, uttered Opinions, Wreaths, Principles, Convictions, moral Tendencies, epileptic Attacks, Childbirth, Insults, Suicides, Noises of discussions (Art-for-art's-sake, Form and Idea), etc.

—Villiers de l'Isle-Adam, explaining some of the settings of his automatic theatrical public, "La machine à gloire," *Contes cruels*, 1874

Terra's blog is titled "Tyler tyler honored with rd Jonas E," and subtitled "Nine State Regulators Investigating Auction Bonds, Group Says. The City of Tyler Traffic Engineering Department installed the City."[48] One of her posts on July 16, 2008, titled "Tyler State Trial Law Litigation Lawyer Attorney Robert M.," opens:

```
Our web servers cannot find the page or file you asked for.

Best choice of the month: _blood pressure.

Button to return to the previous page. Astronomers on verge
of finding Earth's twin. The cost of having a product regis-
tered is now estimated to be around million. Century, the
public may demand that the federal _Bar board massachusetts
overseer_ register products that are effective against bed
bugs.

My mother lives in _Affect metoprolol_ side housing and has
been dealing with this for about a year now.
```

This continues for another 1,300 words, and Terra posted three times on July 16 alone. In June, she posted 160 entries, about five a day, each from several hundred to a few thousand words. This is not her only blog, either; according to her Blogger profile, she has eleven others, with titles like "S first try boosted the team as Biko" and "Only two USB fujitsu three is always." The bizarre stop-start rhythm of her posts makes them difficult to stop quoting. Their language has no heritage in oral speech and lacks the syntactic edges that imply beginning and ending. As with litspam messages, the jolting movement from paragraph to paragraph feels much closer to channel-surfing cable television than to any literary medium: "Oprah ends three weeks of vegan eating. Astronomers on verge of finding Earth's twin. Seeing more people living out of their cars." Then comes a sudden transition into the diaristic, with "I" sentences, opinions, and the rhythmic clauses characteristic of online thinking-aloud: "I don't think it's a numbers game, but I think whatever view you end up with, it does not have to be a majority point of view, that reasons have weight, not just adding up whoever agrees with you." Her posts are full of links, most of which go to similar blogs: vollybllgrl's blog "It was a powerline that brought down a Black Hawk black last night in northeast Alabama," for example, or a post on the blog "Default title" by manning6029 with the oddly Ballardian phrase "Picture of blonde girl in Morocco is new twist in Affect."

Terra is, of course, a robot, as are vollybllgrl, manning6029, Geriut of "The most dazed part of our democracy," etylycigob of "The Triad Lady Knights cross country team had a kylee season," and countless more. They are producing *splogs*, or spam blogs—one of the forms search engine spam took in response to the PageRank strategy of Google and its third-generation search imitators. Splogs now account for more than half of the total number of all blogs. (In comparison, second-generation nonblog spam pages, stuffed with keywords and links, form roughly 8 percent of the total of all web pages being indexed.)[49] The patterns of data from splogs and *spings* (spam pings—a link signal sent by a blog, presented on the linked blog like a comment, and theoretically driving traffic and PageRank) map with striking accuracy onto the patterns of email spam, with similar spikes—around the holidays, for example—and mysterious troughs, during which some waning of the moon causes output to die down for a few days or weeks. How do they work?

As the PageRank system became more widely known and understood, Google gathered market share and other search engines began to follow its more refined ranking model. (Google's ranking system is considerably more elaborate than the bare bones of PageRank, of course, and continues to grow in parameters and complexity to this day, but the basic outline is what search engine spammers were responding to—and that suffices to understand their methods.) A variety of strategies developed as websites with a high PageRank were transformed into kingmakers. A link from them could move a page into the top ten or top three returns of the different search sites, boosting attention and revenue. The theoretical notion of a "reputation economy" was becoming thoroughly applied here. Link trading began as a second-generation approach, along with requests for a positive mention and a clickable link, and third-generation search amplified these methods. Sites issued "Best of the Web" awards, "Top 100 Sites" awards, and so forth; awards included a badge, a little image, and a snippet of code to be copied into the winning site—a snippet that included a link to the award-giving site. The human user saw a little badge image, but the search engine spider saw an outgoing link, that is, an endorsement. New habits of use and etiquette appeared among ordinary users of the web: a comment in a blog post included the commenters' websites along with their names, to rack up another link. Posting something without including a "via" link to the person you got it from—the "via" being an additional outbound link as a kind of thanks for using their discovery—became increasingly rude, the sign of an uncouth person.[50]

These techniques only brushed the surface of the transformation in spam practices. PageRank tried to solve the problem of relevance and the problem of spam in one stroke by incorporating the expression of social relationships, communities, and human choices. In theory, social structures are much harder to game for spam purposes, but their robot-readable expression online is not. The second-generation-based award badges were one strategy among many. *Domain flooding*, for example, was the creation of tens or hundreds of sites to redirect to the target site. Link farms or "mutual admiration societies" arose: these were huge link-heavy blocs of sites, each page linking to many of the others, with their accumulated "votes" rented out. They charged for outbound links from the farm, like penurious aristocrats charging to have their renowned ancient name and reputation associated with some unknown member of the nouveau riche. In the third generation, spam began to move into the creation of its own social graph—producing the appearance, if not the reality, of its own society.

Generating the expression of a nonexistent social phenomenon required the creation of much more content than previous search-spamming projects while avoiding certain telltale signs of robots at work. Old-fashioned attempts to artificially alter the link graph had signature patterns. The bulky shape of heavy cross-linking within a group of sites, all with only a few inbound links (because spam pages are lonesome), created little islands of intense self-endorsement with no outside involvement. To the right analytic tools, it's a pattern as obvious as the newspaper ads taken out by vanity publishing houses for their new releases with the blurbs solely from friends and other writers in the same situation. Search engines could correct for these islands with modifications to the algorithm. Also, although complete web pages could be almost entirely automatically generated, they still required the purchase and maintenance of a stable of domain names and hosting plans with a service provider, which could get expensive. What was needed for third-generation search spam was a way to very rapidly generate new content, seeded with links, in a wide variety of different locations online as in a genuine community.

In 1999, a company called Pyra Labs launched a service called Blogger. The concept of a *weblog*—a chronological series of entries from newest to oldest—is pleasantly intuitive and diarylike; the concept of Blogger, as of so many related systems from Flickr to Wikipedia, was to provide people with an equally intuitive means for publishing their content. It was remotely hosted, so you did not have to own a website domain name or pay for hosting; many of its processes were automated, so you did not have to design it or do any coding behind the scenes; and it had a useful and increasingly sophisticated Application Programming Interface (API). An API is a set of requests that a web service can support from other programs—tools that programs can use to engage with the service. An API makes it easier to automate the publishing process, and on a platform like Blogger (which was acquired by Google in 2003), this automated publishing requires very little effort to manage a lot of content. You can delegate the account creation process, the choice of settings, the ratio of outbound links to content, and the posting frequency to a program. The missing piece here is the words for the blog, but words come ready-made in the form of RSS feeds.

RSS (the acronym originally stood for RDF Site Summary but has changed to mean the more explanatory Really Simple Syndication) is a format closely associated with the development of blogs; it makes new posts or other changes on a site available in forms that are easy to use. Feed readers can gather the latest entries from RSS-enabled sites, material can be forwarded to mobile devices, and a page can feature the headlines or recent posts from other sites. From the perspective of a spam blogger, this feature is like a faucet for words. Samuel Beckett once said of the Cut-Up technique of William Burroughs and Brion Gysin "That's not writing, it's plumbing"—a prescient remark now that we have a means of writing that really is like plumbing: lay the pipes, the tank, the cut-off valves, and then open the taps and leave the room. A splog production system will pull in RSS feeds from other blogs and news sources, chop them up and remix them according to rules, insert relevant links, and post the resulting material, hour after hour and day after day, with minimal human supervision.[51] Terra put a new post up already, while this section was being written, titled "After it became an tyler city": "A witness reports a nun went crazy upon realizing that the man next to her in line was the Epinephrine frontman. Ghost Town Poster Disappoints Me, Gervais. Some things, like gravity, must also be close to." And so on, and on, and on.

All splogs do not sound alike: some are built on the "excerpt" model, with fragments of about 350 characters taken whole from other blogs. These fragments are chosen from posts that are polling particularly well in Google, with good keyword metrics. Their goal is to make money through contextual advertising, in which page views and the occasional clickthrough are the best that can be hoped for. These create parasitic relationships with authors. One of the Internet's many, many interchangeable product-review bloggers notes that being excerpted by splogs is a sign that you are choosing the right topics and words, because the splogs are copying you; if you want to attract more of them, because they offer backlinks to your site with their excerpts, "create posts with a popular keyword, like iPhone for example."[52] The behavior of excerpt splogs is straightforward: they are drawn to the right language like ants to honey.

Splogs built on a full content model, as Terra's is, play a larger and more subtle game, cross-linking in their hundreds and thousands to distort the shape of the web. Each splog is assigned a set of keywords and feeds from which to pull related text. This is why Terra's blog sounds like the product of someone with a feverish, pathological obsessions with Tylers. It pulls a set of RSS feeds and headlines based around "Tyler" as the keyword, with a few others for variation; thus post after post reports from a strange universe where several cities and schools named Tyler, the director Tyler Perry, the economist Tyler Cowen (who blogs), and posts and news articles that mention Tylers are all of equal significance. With experience, one begins to see the patterns. "The TV adaptation of the big screen movie features Nicole Ari Parker, Vanessa Williams and Malinda Williams" refers to one of Perry's projects; "The sociologist Max Weber introduced a distinction between consumer" is a broken fragment from Cowen. Interspersed with this Tyler compulsion is the jarring appearance of the functional language of web design, as in "Button to return to the previous page," used within paragraphs of first-person sentences.

How far removed language is at this point from anything meant for humans! Terra's blog links to other splogs, which link to still more, forming an insular community on a huge range of sites—a kind of PageRank greenhouse that is not in itself meant to be read by people. A human seeing a splog post immediately knows that something is wrong and can flag the splog to be taken down by the network administrator. Splogs of Terra's type are not meant to interact with humans at all; they are created solely for the benefit of search engine spiders. They do not imitate individual humans—they are not the computational equivalent of "George Kaplan," the nonexistent secret agent whose train tickets and hotel rooms in *North by Northwest* are meant to convey a particular human life. They only work from a distance, appearing to be groups of people, with the language and links functioning in aggregate. If splogs are like any previous technological artifact, they are akin to the "QL" sites constructed during the Second World War to mislead night bombing runs: rickety structures of pipe, wooden frames, wire netting, and lights that if seen from far enough away look like a town, with railway signals, lamps, and open doors.[53] Taken in statistical total and algorithmic analysis, splogs resemble the patterns of a thriving community. Their posts are pitched at precisely the level of complexity the spider requires to accept their input as human, and they adapt human text for other machines to read and act on. Influence on humans is a second-order result.[54]

THE QUANTIFIED AUDIENCE

Google, as representative a company of our era as Ford was of the 1910s, is not in the business of search but the business of advertising—its ad services provide 97 percent of its revenue.[55] These ads take the form of little squibs of text or images, often displayed in response to particular search keywords. If a site owner puts some of these ads on his or her web page, they can receive some amount of revenue, generally very small, on a per-impression basis (that is, every time a page with the ad is loaded in a browser) or a per-click basis (a viewer actually clicking on the ad to visit the advertiser's page). Google gets a cut of this revenue as well, and all those served ads on blogs and web pages, sponsored links in search results, and ads accompanying the conversations in Google's Gmail service accumulate into the company's income, and this pays for nearly everything else. (From this fund also comes the oceans of free content whose hosting is paid for out of an individual's share of this money in return for running ads on their site.) So if ads are the business, and content merely the enticement—that is, the ornament on the engine—why not optimize for advertising?

Hence the splogs and spam sites containing post after post and page after page of text automatically gathered and generated to best fit Google's search engine algorithms and filled to the last pixel with advertising, so that every page view and clickthrough is maximized as a source of revenue. The ads on a spam page may be entirely served through Google's affiliate advertising program—in other words, they can be a significant source of revenue for Google. What this means is that search engine spammers running their vast stables of spam blogs and sites are not anomalous. They are making the greatest possible use of the technologies and economies available, constructing a system in which all the extraneous matter of people and conversation has been pruned away in favor of the automation of content production, search results, clicks, and ads served. (The "Enterprise" package of one of the many businesses in this field will mass produce up to 1,000 blogs for the subscriber, turning out 10,000 posts a day around the 150 keywords of the subscriber's choice—a daily volume of text that quantitatively dwarfs that of entire literate cultures and historical epochs.)[56] This system in turn puts Google in the contradictory position of having to analyze and expel many of their most dedicated customers: those who deliberately overexploit, and accidentally overexpose, the financial and attention economies and technologies that underlie the contemporary web.[57]

The mass production systems for human-authored text known as "content farms" exacerbate this contradictory role. Demand Media, an exemplary case, commissions content from human writers (who are willing to meet very low standards at high speeds for very little money) on the basis of an algorithm that determines ad revenue over the lifetime of any given article. It then posts this content through dozens of domains such as eHow.com and Livestrong.com. Generating, at peak, thousands of articles a day, Demand Media can create a simulacrum of knowledge convincing enough to attract both search engine returns and the clicks of actual humans (despite producing a kind of nonsensical poetry of uselessness, the correlative of spam's machine-mangled posthuman semantics, with articles such as "How to wear a sweater vest" and lengthy reviews of deodorant containers). As C. W. Anderson observes, content farms are engaged in the attraction and manipulation of a "quantified audience," a strategy that marks a nebulous border space between more reputable and legitimate media production and spam as such.[58] After all, these are very precisely targeted articles written by people for people; at what point do they cross over from the space of a merely frivolous or attention-grabbing article that a newspaper would run and into the domain of network misbehavior? When does algorithmic quantification part ways with the canny editor who knows that sex, serial killers, and how-to stories sell?

Throughout this history, the spam has produced definitional problems. Easy as it might be to identify a canonical example—one of those laughably awful filter-beating projects described earlier, for instance, with Cialis links interpolated into a lexical pulp made from the Federalist Papers—it's the edge cases that are problematic. Whether we're talking about free speech on Usenet, the policy questions of legitimate marketing and commercial activity conducted over email, or the desirable but spam-ish messages that trip the filters and disappear, there is always friction not around the most egregious case (no one argues for Leo Kuvayev's "\/1@gR/-\" messages) but at the blurry places where spam threatens to blend into acceptable use, and fighting one might have a deleterious effect on the other. The realm of "social spam" and the quantified audience is the blurriest of these—where fairly acceptable and established methods of getting attention and audience management may begin to shade into spam.

"The algorithm is fed inputs from three sources: Search terms (popular terms from more than 100 sources comprising 2 billion searches a day), the ad market (a snapshot of which keywords are sought after and how much they are fetching), and the competition (what's online already and where a term ranks in search results)."[59] This statement could be describing an extremely well-run empire of splogs, but it is journalist Daniel Roth's description of the Demand Media's operation. The algorithm outputs what's going to be profitable, the jobs are posted to a separate site to find labor, and then a person writes the entry. "It is a database of human needs," Roth adds, but that isn't exactly true. It is a constantly updated collection of those queries whose results can consistently make some money over time; "needs" is rather too grand. Anderson dissects the normative commitments this kind of "algorithmic journalism" makes—because it does indeed make commitments and reflect beliefs, which we should not dismiss too quickly, facile and self-serving as those beliefs may seem. They are where different constituencies stake out territory and make their arguments in the technological drama.[60]

Anderson identifies five commitments, in which we can find a pronounced echo of the lineage of spam from which content farms and other types of algorithmic journalism partake.[61] It's built around "big data," and it thus features the blurring we've seen before between human and machine input and judgment. It takes as cardinal the idea of "consumer choice"—after all, it's "demand" media that can claim to be giving those query-typing people *exactly* what they want, to a very high degree of mathematical precision, and therefore without any pretense of paternalistically filtering or "curating" the information for their benefit. Finally, it is future-oriented, because it is predictive—rather than reporting the news of the immediate past, it can look to what's trending and have content produced to that apogee, like a Wienerian cybernetic gun shooting to where the aircraft will be when the round arrives. Looking at these five beliefs embodied in the content farm project, we can see again the capture of relevance, in a more refined form than before. Rather than send a million emails in hope of a handful of responses, make a million articles that will be perceived as relevant enough by the search engines to get top billing for a handful of searches and by a person to click through and contribute ad revenue.

For a slightly different approach to this project, we can turn to AOL, where the historical ironies become almost rich. The company's walled-garden approach brought huge numbers of new users to networked computing in the United States. Not coincidentally, it also brought the enormous waves of inexperienced users that kept breaking the mores of "netiquette" and provided a source of profitable chum for the early spammers. It is now reinventing itself as a frantically SEO-gaming content empire. A leaked internal memo concerning "The AOL Way" revealed a fascinating project to use a tightly coordinated human staff to generate an enormous amount of text against which to serve advertising.[62] The number of articles produced is to jump from 33,000 a month to 55,000—five to ten articles produced per staffer per day—built around a system of metrics based on "Traffic Potential," "Revenue/Profit," "Turnaround Time," and "Editorial Integrity," with point-by-point questions such as "What CPM will this content earn?" ("CPM" means Cost Per a thousand [thousand = M, as in Roman numerals] times that an ad is loaded—that is, the anticipated revenue to AOL.) "Is this story SEO-winning for in-demand terms?" appears in the checklist: how many vitally popular keywords can be used here to net the greatest number of searches? In this light, AOL's purchase of the massive content producing and aggregating site the Huffington Post fits the model of algorithmic journalism Anderson describes. AOL is not buying a popular and potentially problematic cultural property, like Sony purchasing a movie studio or Condé Nast losing money on the *New Yorker*. They are buying a factory—an assembly line system with a proven track record and a well-managed, if grueling, schedule, that can produce or aggregate material appropriate to trending topics reliably enough to generate page views with the latest in top-ten lists and the travails of reality TV's stable.

Is this spam? Not precisely, though that term is often applied—but "spam" has never been precise. There is something in the disposable and opportunistic nature of the material produced, and the mingling of automated and human infrastructure used to produce it, that seems similar—a cynical project to monopolize the conversation and commandeer the space of relevant information. *Linkbait* is a related term, one that originated in the SEO community to describe the strategy of producing relevant, highly "linkable" content in hopes of drawing traffic, and therefore advertising revenue, from "link-savvy bloggers and web content creators" and the "hundreds of sheep-like content creators" who follow them (to quote one of the earliest appearances of the term in the fall of 2005).[63] Having originated as a positive phrase for an exploitative strategy built on the kind of lightweight trend-based content long popular in the magazine industry, "linkbait" was soon adopted as a negative descriptor that covered the same content. From the perspective of readers looking for something of depth, it was the perfect term for the vast algal blooms of linked content with catchy titles, top-ten lists about trending topics, wild claims, and needlessly contrarian stances, all delivered with only a few hundred words per article. The term has now spread from its SEO roots to describe other cultural phenomena perceived to sacrifice argument and evidence in favor of attracting notice.

Consider yet another version of this idea, as applied to individual self-promotion: *personality spamming*, a term coined by writer Merlin Mann as a slightly bitter joke about the use of microblogging service Twitter. Personality spamming is the work of arrogating attention for oneself, using social media to build an audience—often a very carefully quantified audience of "followers" and "rebloggers"—rather than a network of friends, as was the initial, notional promise. It is a witty condemnation of the socially acceptable but aggressively eyeball-hungry work of those who would be, or act like, celebrities, "influencers," or "thought leaders." The top reason for unfriending on Facebook is "frequent, unimportant posts," and many computer-based clients for Twitter have a "mute" feature so that you can ignore messages from some users without having to unfollow them and then refollow them later (which would let them know you had turned them off for a time)—indicators of personality spam as a feature of daily life.[64] As Anderson suggests with algorithmic journalism, these practices reflect something genuinely new, and as yet not clearly theorized, distinct equally from Habermasian communal conversation-as-deliberation as from the blandly managerial product, shaped by layers of human talent for the broadest possible distribution, of Adorno and Horkheimer's *Kulturindustrie*. There is a reformulation under way in which the question of acceptable modes of social expression and self-promotion are being weighed. Drawing on Alice Marwick's research, we can see some of these new modes being forged by individuals who turn themselves into the epigones of major advertising and marketing firms: the brand is you, the goal is to accumulate relevance for certain terms or ideas so that you can become, in some nebulous sense, "influential." (From that state will come the book contract, the speaking fees, and the TV deal, presumably.) Thus the method is to treat every platform, gathering, and interaction as a marketing opportunity to configure oneself and one's activities to suit the search algorithms.[65]

IN YOUR OWN WORDS: SPAMMING AND HUMAN-MACHINE COLLABORATIONS

The gradual predominance of the algorithm in the project of spamming appears in the filters and the spam created in response to them, in search engines and their manipulators, and, as will be shown, in the grand global project of the botnets. However, it is most eerily seen in those places where algorithmic initiatives and human labor intersect. Content farming is a great instance of this combination, but there are other examples, some still more intimate, where human and machine production are meshed to beat the automated security of antispam systems. Mechanical Turk, for instance, is a truly strange and contemporary thing: a marketplace for *crowdsourcing* small units of work that can be done by a person on a computer. Under the rubric of "artificial artificial intelligence," it's a venue in which a "requester" (in the Mechanical Turk terminology) with a task can break it up into fragments called *human intelligence tasks* (HITs), offer a price per task, and then see if any of the cloud of "providers"—workers looking to pick up some small quantity of micropayment labor, akin to the "content producers" waiting for new jobs from Demand Media—will take them up. Amazon's system coordinates the workers, the task fragments, and the payments. (If you had a forty-five-minute interview in an mp3 file, you could break the audio up into two- or three-minute segments, upload them to Mechanical Turk, offer a dollar for each transcribed segment, go have lunch, and return to find much of the work done.) The service is estimated to have 100,000 workers in 100 countries, with the majority in India and the United States. It gets used for transcription work, as in our example, as well as in database projects, surveys, image tagging, and more recondite activities. It features a range of HITs for rewriting texts of various lengths, many of which appear to be for services that provide plagiarized or "pre-written" essays and papers to paying students—the rewriting of the texts ("in your own words") makes them harder for a teacher to identify with a Google search.

Simply creating HITs to send out spam email would be pointlessly difficult and expensive compared to easy and massively automated processes such as botnets.[66] But the Mechanical Turk system is ideal for engaging in social network spam.[67] ("Social networks": of course, all networks are already social, regardless of whether they want to be.) Many sites now come with built-in models of user action and selection, from voting to public bookmarking to collaborative filtering, providing different ways for the group to assign salience and value. Aside from the direct benefits of traffic from one of these sites, as users see an interesting link and click through on it, getting linked on a major social networking site is a good way to boost one's PageRank and get better search returns. The now decades-long quest of the search engine spammers to move up in the search rankings has thus migrated into the new territory of social recommendation systems. "Could you please bookmark my site / Using one of the following sites: http://www.del.icio.us/ http://www.stumbleupon.com/ http://www.furl.com," says one Mechanical Turk requester, offering a rate of $1.75 per bookmarking.[68] Suddenly, in the eyes of the algorithms of the social networking sites and search engines, there is a rise of interest on the part of reputable and high-value real-human social site users in this ad-laden website about mortgage restructuring or celebrity sex tapes.

Craigslist, meanwhile, offers a very different challenge and reward for those who would spam social networks—a challenge that has led to a strange human-machine arms race. Craigslist is a free site for posting classified ads, from bicycles for sale to apartments for rent (and great volumes of personals and "missed connections," a mass index of urban loneliness and yearning). It offers free space for ads on a site with hundreds of local city instances, sitting ninth place in the number of page views served in the United States (as of this writing), up there with Google and its properties Wikipedia and Facebook. Craigslist therefore obviously needs to protect itself against spammers. One of the characteristics of spam is duplication of text—it's one of the properties that Bayesian filtering seized on as a weak point—so Craigslist blocked multiple ad postings with the same text or from the same network address. They required a valid email address to post, emailing a confirmation demand to that address that had to be clicked before the ad would be posted. They used a CAPTCHA system—the deformed letters on weird backgrounds that only humans can read, in theory, to verify their nonbot status—to block automated posting tools. Finally, they allowed other users to flag an ad as spam so that the site's moderators could delete it. The spammers, in return, developed tools such as CL Auto Posting Tool and Craigslist Bot Pro 1 (the banality of the business of spam: $67, Windows only, "allows you to automate your personal and business online advertising") to sidestep each of Craigslist's defenses. Textual polymorphism—individual variations in the language of a spam message—could defeat the duplicate message detector, just as it does in email. Proxies could be used to post ads from lots of different network addresses, with valid email addresses for confirmation messages stamped out like license plates by programs like Jiffy Gmail Creator. Captcha King can fill in the CAPTCHAs. Monitors were developed to detect when an ad was flagged as spam so that it could be automatically resubmitted.

Craigslist then turned to telephone verification. To post an ad in certain categories, you have to take an automated phone call or text message with your confirmation password before the ad will go up, with only one ad per phone number. The spammers tried using voice-over-Internet (VoIP) services such as Skype, which in some cases made it possible to generate new phone numbers. Craigslist blocked those. "My assumption is probably accurate that CL is looking at the national database that distinguishes which numbers are voip and which arn't [sic]," wrote one spammer in an extensive technical discussion devoted to overcoming these new developments.[69] The spammers turned to services that could allow them to register additional phone numbers for a small fee. Craigslist blocked those, too.

The spammers turned to other platforms: "why don't you guys take a laptop and go to: truck stops airports bus stations you should find close to 100 pay phones there"—and use the phones and their numbers for the verification messages.[70] Another spammer reported back: "I used to have 140 accounts all done by me at payphones. It took me about 3 days. It was not easy and it was boring." A more ingenious, almost Mechanical Turk–like distribution of labor project followed as the culmination of these efforts: "some are creating pages of ringtones [for mobile phones], if a person wants a ringtone All you have to do is to receive an sms (craigslist) in her cell and received this code placed on the website and can automatically download your ringtone." In other words, people with mobile phones in search of free ringtones will act as the distributed phone verification system to compensate for Craigslist's antispam initiative: a random voluntary population, organized remotely by machine, helping advertisers to swamp the community platform without ever realizing that they're doing it.

CAPTCHAs, that border between the human and the robot-readable used by Craigslist among many other sites and platforms, have long dogged spam production, making it harder to start new Blogger blogs or open more free email accounts, and spammers have been working assiduously on different fronts to overcome them. In May 2008, a truly odd breakthrough took place. Security company Websense documented a series of attacks on the account-creation process of email services. Many requests for accounts kept hitting the CAPTCHA stage, and most, but not all, failed.[71] There was Russian-language evidence of offers to pay small sums for the solutions to CAPTCHAs, but the pace (replies in six seconds) and the failure rate (nine to one) suggested that computers were doing the solving. ("We still believe there is human involvement," said the company's statement.) Later, Websense also documented a significantly improved CAPTCHA-cracker, one to which the spammer's computers could pass their CAPTCHA problems as they made new email accounts. This program could take the image of distorted text and return a result within twenty to twenty-five seconds, with a significantly improved error rate of one success in five to eight tries or between 12 percent and 20 percent—not bad at all.[72] Botnets, with all their spare computing power, are ideal for brute-force attacks on the computationally onerous processing required to analyze CAPTCHAs.

At the same time, the opposite tack is being taken by services such as Captcha King, mentioned earlier in the Craigslist-spammer arms race, that advertise a series of aristocracy-themed payment plans (Royal, Imperial, and Emperor) for CAPTCHA solutions sold in batches of thousands. Their method, which integrates with spamming software like automated Craigslist posting engines, Jiffy Gmail Creator, and MySpace bots, retrieves the CAPTCHA images "for manual entry." An outsourced staff sits there all day banging out CAPTCHAs, with a guaranteed "success rate of 95% with a response time of less than 90 seconds." Those poor souls, whose work makes regular data entry look exceedingly pleasant by comparison, are essentially being paid to be human, that is, to exhibit a theoretically solely human characteristic. (Another service along the same lines, KolotiBablo, tells us with its pay rates that "bare humanity" isn't worth much in itself: between US$0.35 to US$1.00 for every thousand CAPTCHAs solved—meaning a bit under $3 a day for eight solid hours of typing in CAPTCHA texts six times a minute.)[73] In their work, and in the statement "We still believe there is human involvement," we can hear the echo of Alan Turing's clattering teletype in the parlor playing the Imitation Game. Are the CAPTCHAs being solved by a distributed workforce of deeply bored humans or by increasingly sophisticated optical character recognition programs running on a network of compromised machines? Some details can help distinguish them, but the fact that the two can be intermingled and difficult to identify—who's on the other end of the line?—calls up the essential problem of Turing's thought experiment. As Kevin Kelly put it, "What if spammers come up with an artificial intelligence before Google does?"[74]

In response, the kind of technology used to tell computers and humans apart is also being pushed to greater sophistication—yet another arms race. Work is now being done on presenting moving pictures (such as a galloping horse) made of animated blotches against a blotchy background. It's the sort of thing that a human can identify but a computer would find exceedingly difficult, at least so far.[75] There is an inventive world of vernacular bot-stopping solutions on personal websites: an email address ending with "oryx," with a note to remove the "genus of antelope" before sending; a very simple joke for which you must choose the obviously correct punchline; a photograph you must briefly describe ("am I in the house or on the beach?") before you can send a message—the kind of tasks that are trivial for humans but require inference impossible for the crude programs currently being sent to gather addresses and post comment spam. Intriguingly, one of the CAPTCHA-busting sweatshops described earlier, a Russian service called Antigate, keeps Westerners at bay by requiring visitors to enter the name of the current Russian prime minister using the Cyrillic alphabet, a "culturally restricted CAPTCHA," meant to not simply fend off bots but to sort out groups of humans.[76] The territory of what is uniquely and reliably human (and can be automatically tested at scale, over different kinds of interfaces) is one of interesting zones given to future technologists to explore—if only to keep the botnets at bay.

THE BOTNETS

"By now I don't know exactly what there is in the worm," announces the protagonist. "More bits are being added automatically as it works its way to places I never dared guess existed." He continues, "And—no, it can't be killed. It's indefinitely self-perpetuating so long as the net exists. Even if one segment of it is inactivated, a counterpart of the missing portion will remain in store at some other station and the worm will automatically subdivide and send a duplicate head to collect the spare groups and restore them to their proper place."[77] This passage from John Brunner's 1975 science fiction novel *The Shockwave Rider* is quoted at the beginning of John Shoch and Jon Hupp's remarkable 1982 paper "The 'Worm' Programs—Early Experience with a Distributed Computation."[78] It is through them, and their work at Xerox PARC, that the worm makes its conceptual, etymological way from Brunner's novel to email spamming's mutations in the new millennium.

Shoch and Hupp were envisioning something quite inventive, particularly for the time: a "distributed computation," that is, a single program operating across many machines and taking advantage of idle processing power to do its work. This "worm" is the first monster from which the others spring with the same essential DNA, the worm that grows at night ("affinity for nighttime exploration led one researcher to describe these as 'vampire programs'") as it segments individual underused machines for a collective purpose.[79] The essential project remains the same, from Brunner's novel through the lab in 1982 to the present day: turning all the little boxes into one big machine. "Instead of viewing this environment as 100 independent machines connected to a network, we thought of it as a 100-element multiprocessor, in search of a program to run."[80] Worms since then have had a long and storied history in legitimate computer science, but the idea of a worm program as articulated by Brunner, Shoch, and

Hupp has also found an extraordinary life in botnets and their spam-financed operations.

Imagine an office cubicle in a big building somewhere in the world—it could be the United States, Taiwan, Germany, or Brazil. The fluorescent lights hum overhead in the drop ceiling. An employee is away from his desk. His computer is playing a screensaver of family photographs. The computer—a standard-issue black Chinese-manufactured clone machine running Windows XP—is idle, but still engaging in automatic behavior over its broadband connection. It checks for new email at the server every few minutes, for example. A small but regular trickle of requests and replies move over its always-on connection to the network.

At some point in the past, perhaps while the computer's user was visiting a malicious web page, downloading and installing a program, or opening an ecard from a stranger, this computer was infected with a bit of *malware*, a program designed to exploit computers. In this case, the malware was a *worm*, a much-developed heir to the Shoch and Hupp worm concept in the form of a parasitic program capable of operating on its own. (This behavior distinguishes it from a virus, which needs to operate inside another program already present on the computer.) Far below any level our employee would ever notice, somewhere in the recesses of disk space, the worm uses spare processing power on the computer and the extra bandwidth of the always-on connection to do its work, turning the computer into a remotely controlled tool for the worm's programmer—and, automatically, into a tool for spreading the worm to other computers. This malware's point of infection can be exceedingly simple and subtle. Perhaps the employee received an email from a coworker's address, warning of a failed message, giving the innocuous and puzzling explanation "The message contains Unicode characters and has been sent as a binary attachment." He downloaded and opened the attachment to see only a page of meaningless symbols. He closed the page, perhaps sent a reply to his coworker—"Had trouble with your last message?"—or ignored the whole event as a computer mystery.

When he opens that attachment, the employee launches the worm to do its covert work. Having installed itself on the computer, it begins searching the host's files for email addresses, to which it sends versions of the infection message, randomly drawing the header, body text, and attachment name from a small collection, all equally puzzling and dull. It looks for the popular file-sharing program Kazaa (one in the group of popular peer-to-peer programs for sharing media files that included Napster, Gnutella, and Morpheus); if it finds it, it copies a version of itself to the directory of shared files under one of several names such as strip-girl-2.0bdcom_patches.bat, office_crack.exe, or winamp5.[81] Now, in the huge mesh of file-sharing computers, someone browsing this user's files—or searching for a "cracked" (free, unprotected) copy of Microsoft Office or a stripping girl—will find one of these files, download it, launch it, see the page of meaningless symbols or an error message, and be similarly, quietly infected. But the worm has much more to do beyond replicate itself.

It also opens a "backdoor" to the infected computer that allows it to communicate with its controller and execute commands on its behalf, turning the computer into a "bot" or "zombie" machine. It begins quietly shipping information back and forth over the available capacity of its connection to the Internet. It checks in with the "command-and-control" channel, on which it receives its instructions from the botmaster. (This channel is often set up using an ancient, robust chat protocol called Internet Relay Chat [IRC].) The instructions given to it are generally along these lines: take this text ("Your Online Banking is Blocked! / We recently reviewed your account, and suspect that your Bank Of America account may have been accessed by an unauthorized third party") and send it as an email to this list of addresses. The computer on the desk in the office cubicle has become a spam-distribution machine and has the capacity to do much more. It has joined the botnet.

Why the "bot" in botnet? Bots are simply programs that can do what they're programmed to without constant human involvement. They can correlate data, hang out in a chat channel providing the rules of conduct when anyone asks, or search the web for email addresses while their programmers are occupied elsewhere. These abilities make them ideal for an enormous variety of computer tasks—and among them is spamming. Far back in the history of online socializing, "floodbots" would join a channel and fill it "with garbage text, endlessly repeated insults, or random billowing storm clouds of data," killing the normal conversation.[82] In 1996, with spam as a targeted marketing model taking off and NANAE forming, a company called GlobalMedia Design released RoverBot, one of the early address-harvesting bots, which would take keywords, find related pages, and search those pages for email addresses so that you could generate address lists related to "real estate" or "manga." And, portending the rise of increasingly autonomous spam operations, there was the spambot ActiveAgent, a little nightmare that crawled web pages looking for addresses and emailing them with preprogrammed text; the author, "Robert Returned," would sell the code for ActiveAgent for $100 to anyone who asked.[83] Of course, there were already more efficient means of amassing and mailing to addresses being developed—methods that would culminate in the botnet.

Our fictional employee's desktop computer is hosting a real worm: first released in early 2004, it was referred to in the security community as Mydoom, and it had good archetypal characteristics for explaining the basics of a botnet. In particular, it has the sting in its tail that brings botnets into conversation with the military. "On the first of February 2004," the worm tells the infected computer, "request the website of SCO Inc., http://www.sco.com, every millisecond, and continue until the twelfth of the month." You request a site when you type "www.sco .com" into your browser's address bar and hit return or click on a link to sco.com: the request is sent to the server at that address, and data from the server is received and displayed on your screen. This is the normal business of servers, and they are built and configured to handle a certain number of requests for a certain amount of data from a certain number of users, depending on resources and anticipated use. If too many requests arrive in too short a time, the server cannot deal with the new requests and the site cannot be accessed—it becomes unusably slow or entirely fails to respond, leaving the user with an error page ("The server may be unavailable," "The server has timed out," and so on). This is called a *denial of service* (DoS). A DoS is often the result of a sudden burst of popularity, when a personal site that normally receives a few hundred visitors a day appears on a major blog or social news site and then suddenly receives tens of thousands of visitors and gets overwhelmed. Such an event can also be undertaken maliciously. It is what the charivari of outraged Usenet denizens did to Portal and Internet Direct as vengeance, swamping the servers with furious mail and big, capacity-consuming image files.

What this command issued by the Mydoom worm meant to do was create a vast phantom population of users requesting the site again and again and again from many thousands of computers all over the world, effectively knocking the site offline for twelve full days, rendering them unable to do business and acting as a devastating blow to their reputation as a company that provides secure servers for enterprise clients. A coordinated action from a botnet, a global network of machines, to take down a website or a server is called a *distributed denial of service* (DDoS) attack. Such an attack can be used to extort money from online companies (such as casinos) by preventing customers from reaching them, to eliminate security firms or other enemies, and to attack civil and governmental Internet infrastructure: it's a transition from tool to weapon, with spam becoming a mere platform for further developments.[84]

The Mydoom worm contained a poignant message embedded in the code: "(s y n c—1 . ★ ★ o ★ 0 1 ; a n d y ★ I ' m j u s t d o i n g m y k ★ ★ ★ ★ o b, n o t h ★ p e r s o n a l ★ ★ ★ ★ ★ } r r y) B G @", usually transliterated as "(sync-1.01; andy; I'm just doing my job, nothing personal, sorry)."[85] The author, or authors, of Mydoom have never been caught; the "job" and "Andy" remain mysteries, known only to a small group of collaborators, competitors, enemies, and friends. This private message from one person to another embedded in the code creates a dizzying sense of parallax in context of the scale of the botnet—a system that makes spam production literally the size of the planet. All those individual desktop and laptop computers in homes, businesses, dorm rooms, and Internet cafés can be seen as a single resource, part of one continuous landscape, and a huge untapped well of spare system cycles, bandwidth, and sensitive information. Once you have the distributed power of many infected computers that are autonomously infecting others in turn, new projects and possibilities arise. A botnet becomes a platform, with spam just one "program" among others that runs on the platform alongside things such as key cracking (breaking passwords and encryption), clickfraud (automated "clicking" on ads to increase revenue to the ad host), identity theft of all kinds, and DDoS attacks—and potentially much more. It is the beginning of a new scale of operations.

THE MARKETPLACE

Life as an apprentice botmaster: the worm you wrote, or more likely bought or stole from a more skilled programmer, has succeeded, proliferating steadily for several days. You now have ten or fifteen thousand compromised computers under your notional control. Their number varies from day to day: perhaps a new infection boom has added a few thousand more, or a patch has been released that fixes the security flaw you were taking advantage of (but only several hundred of the users of your infected machines know to install it, so you do not lose that many bots). People go on vacation, leaving their computers off for a week or two; companies upgrade, and the old machines—your machines—go out to the recycling bin to be palletized and shipped to Accra or Guiyu. Other worm writers and botmasters create programs designed to take over machines and knock off the infections already present, like yours. From day to day, users of infected machines all over the world power them up or down on cycles of nights, weekends, and lunch breaks. The bot population is shifting and unreliable, and you face the very real problem of making use of all of this distributed computing power you have accumulated. You have what security analysts call a "victim cloud" with which to make money generating spam, among other jobs. How do you control it?

On the most abstract level, your method is this: you use the archaic but reliable protocol for real-time messaging online, good old IRC. IRC has a long history of automated interactions in which chatbots have been responding to commands and relaying messages long before the arrival of more sophisticated technologies. All of your infected computers subscribe to your IRC channel, referred to as the *command-and-control* (C&C) channel, and you can easily send out instructions to their population, such as message text and address "target lists" for spam campaigns. This relatively simple arrangement creates another problem, however: now your network of compromised machines has a single point of control, that channel, and thus is vulnerable to attack and seizure, whether by law enforcement and "white hat" good-guy hackers or other botmasters, who could commandeer your channel and use it to make your machines work for them. (Other botmasters trying to take over your network is the biggest ongoing problem you face.) There are ways to make your C&C channel more secure. Perhaps you have managed to obfuscate or encrypt some of the critical traffic and code, such as the authentication passwords you use to control your bots. This trick will keep the other botmasters at bay, for now. The next critical question: how are you going to make money?

As with the development of spam itself, this is all about taking advantage of new affordances: you are on these computers now, and you control them. First you snoop, scrubbing the compromised computers for usernames, passwords, email contacts, financial information, secrets—and you monitor their network traffic for similarly useful material, like anything associated with the keywords "paypal" and "paypal.com," which might have a password attached.[86] (When the security company Finjan seized a server being used to store botnet data, they found 1.4 gigabytes of material from compromised machines in the United States, the European Union, India, Canada, and Turkey, including patient data from health-care providers as well as the usual acres of business databases and email logs.)[87] It's possible to monetize many of these resources yourself, but it's also often time-consuming and potentially dangerous without the right skills—and getting money out of bank accounts and credit cards safely is a very different matter from simply getting credit card numbers and account login information.

Instead, you bring your data into the thriving underground economy that has formed around online crime. You join yet another IRC channel: a screen of names, or "nicks," working out deals in the typo-ridden lowercase that acts as the argot of the marketplace. "i need 1 mastercard i give 1 linux hacked root" pops up; "i have verified paypal accounts with good balance . . . and i can cashout paypals."[88] Trustworthy users who have proved their reliability to the channel's administrators have a +v symbol at the end of their nick, so you know you can do business with them—they are not thieves, "rippers"—at least not among their own. ("report ripperz to @s -Trade OPEN rippers are not alowed [sic] here . . . if u find one show the log.") There are several different ways you can make money at this point. You can sell the data you have stolen from the infected computers under your control to a "cashier," someone who knows how to convert financial authentication information into money. Your cashier may themselves have to work with a "confirmer," who can pose as the sender in a money transfer using a stolen account.[89] (Because cashiers often need to be country and gender specific—a bank will not clean out an account belonging to a female name in Texas if a male voice with a Slavic accent is on the line—an odd economy in, say, "fml CA US UK cashout" cashiers has developed.) You can also try cutting a deal with the cashier to keep more of the profit.

You can sell your botnet as a whole for a smaller but quick profit: the going rate is between four cents and a dime per compromised computer.[90] They pay you the total, you send them the passwords and other information for the C&C channel for the bots—the keys to the spam factory. You can also rent time and capacity on your botnet for all the services it can provide: hosting cracked software for download, hosting fake sites for phishing messages (where people can input their passwords, in response to an email, under the impression that it belongs to Facebook or their bank), delivering DDoS attacks, and running spam campaigns. The channel is a great place to get set for your own spam projects, as well, with databases of email accounts, including "targeted" collections—for example, those professionals with bank accounts more likely to fall for a bank-message phishing scam—there for purchase and trade. You can get lists of netblocks (ranges of Internet addresses) that are notably vulnerable or heavily monitored or that belong to certain organizations that you might want to take advantage of or avoid. Finally, you can barter for all of these things, transacting any one for any other: time on your machines in return for a list of addresses, some credit card data in return for a few thousand more machines for your network. After a good spam campaign, with a mix of pharmaceutical messages for a client, paid for in batches of a million and sent to a cheap, inferior list of addresses—and phishing messages for your personal profit, sent to a more precise, targeted list—you can come back to the market with more data to sell, and more money with which to buy work and data from the others.

The market is transnationally hopping—though it looks, like so much of your working life as a global criminal, like a window on your screen with text in it. People take advantage of the primitive text/background color choices to make their offers stand out in a visual shouting match of green text on brown or white on electric blue. A variety of typo-ridden languages are in use. On an average afternoon, somebody with the nick "TOrPedO`" tries to drum up business: "CA (DOB + mmn + SIN + ATM PIN + Paypal with email access + Drivers License) = 12 $—AU (DOB + mmn + Paypal with email access + Drivers License + Medicare card number + ATM PIN) = 10 $—Also EU fulls selected countries could be spammed on Request. . . . SELLING cvv2s Available for Sale: Cvv2's US bundle of 20 for 60$—EU countries bundle of 20 for 75$ ... SELLING MAIL LISTS 1Available for Sale: US, UK, CA, AU, European: IT, ES, GR, FR, GY. Bundle of 5mb = 40$—PM me now."[91] "PM" is "private message": step out of the public space and get the deal done.

If TOrPedO` is you in this scenario, you can move spam "on Request," you have lists of addresses targeted by country for sale to other spammers, you have all the identity theft basics at $12 a pop, as well as bundles of CVV2s—the three-digit Card Verification Values used to confirm credit card transactions when the card is not physically present—priced to move. Some of the data you have accumulated needs to be turned into money, and the nick PhuckedUp is looking for clients: "Legit PinCashier, Looking for Supliers, i cashout FCU, CU, Small Banks, with limit of 3k ! msg me only serious supliers !"—FCU and CU being "credit unions," that is, smaller banking operations. You have a lot of competitors in this business. zgfrik posts: "selling abbey [Abbey banking] account with 23k on it,price 1000$—msg me if interested." As in markets everywhere, trust is a problem, and the warnings fly: "BOSNIAN RIPPERS Ognjen Miric AND Ervin Residbegovic—BOTH LIVES IN Bosnia And Herzegowina! Sarajevo! ZIP: 71000 DONT BUY FROM ANYONE FROM BOSNIA // Sarajevo! YOU WILL LOSE YOUR MONEY 110%!"

You post your notice: "=(REAL BANK LOGINS SPAM SUPPLYS)=(SELL BANK LOGINS\PRICE DEPENDS ON BALANCE 10% FROM IT)=(BIG BASE!)=(ADD ME>," followed by a chat name and email address. Later, as you meet others in this world, you will move on to covert password-protected channels where more serious action happens. You have joined the twenty-first-century spam economy.

It's not a bad living, as documented by analyses of Russian forums devoted to doing malware, spamming, and credit card theft deals.[92] A million spam messages sent on behalf of a client costs the equivalent of a hundred U.S. dollars—and there's a bulk discount, of course. A million addresses for $120, more if you want them sorted by country. Fifteen dollars for a hour of denial of service attacks; more for a more sustained attack, which requires more cunning to outwit the blocking strategies the target might employ as they catch on. Given how much it can cost a target to be down during the attack, it's a great way to make money by extortion. You can sell a malware program called "Pinch" that searches for banking data and passwords from infiltrated computers, and you can also sell the raw data you acquire—$10 a megabyte, for others to pan through in search of profitable information and to go to the additional trouble of actually extracting the money. (The transactions between parties in the business are done through services like Yandex and WebMoney, services akin to PayPal but with greater market penetration in Russia and Eastern Europe.) If you buy a hundred "good" credit card numbers (verified, with CVV and all the ID information, with high spending limits) for $10.66 apiece, of which perhaps half can actually be used buy and ship goods to Russia for resale or fencing before you set off their antifraud detection systems, that can still produce a few hundred dollars' worth of value per card, for a profit of $13,000. Not bad at all.

Still better are advance-fee fraud messages—the "Nigerian spam" described earlier—at a cost of $20 for 200,000 messages (they're more expensive because they have to be more targeted in the sending and tailored in the writing, themed with recent news and somewhat plausible details), with a response rate of 2 or 3 percent and an average take of $1,922.99 per victim. Even if they spammers don't net a really big fish, they can ultimately expect to clear about $200,000 in profit, though it is a lot more work. There may not be honor among thieves, but there is good customer service. The interdependent parts of this economy include agreed-upon systems for testing product (a sector of a botnet to confirm the available bandwidth, a few credit cards from a batch to make sure they're real and to check the balances), money-back guarantees, nicely designed interfaces, partnership programs, and, charmingly, free champagne for closing a deal together. As Holt argues, it makes sense in the short term to lease a botnet rather than build one of your own—you can send spam and do attacks with a somewhat higher profit margin and no maintenance. But what if you are a truly gifted and visionary programmer? What if you want to build a better botnet?

INSIDE THE LIBRARY OF BABEL: THE STORM WORM

Paul Graham had speculated that "the spam of the future," designed to better beat filters, would take the form of squibs of text and a single link: "Hey there. Thought you should check out the following: http://www.27meg.com/foo." It looks innocuous enough, but there was an unforeseen element that could be added to this mix, particularly after September 11, 2001, and similar shocks: news, or the promise of news. Messages in a new spam vernacular began to arrive, promising dreadful events and scandal, from celebrity gossip ("Justin Timberlake Says 'Britney Shaved Her Head Bald For Me,'" "Will Smith found dead in bathtub") to the amusingly bizarre ("Bigfoot found, shot down in cold blood") or the politically startling ("Chinese missile shot down USA aircraft"). The subject line promised much, with brief body text ("A hunter claims he saw the legendary beast known as Bigfoot") and a link to the story. The link directed to a web page or a download, the point of malware infection, just as the purported file sent by the coworker did in the Mydoom instance. Such a message is classified as a *self-propagation* spam campaign in the language of the antispam community—spam to add more machines to a network. In early 2007, the self-propagation message that dominated the field was "230 dead as storm batters Europe": the vehicle for the eponymous Storm Worm.

Storm spread swiftly, but more worrisome and fascinating than its speed was the technical muscle behind the scenes, visible to those antispam groups and security companies that watched the bot world. It began simply, albeit at a higher level than the primitive second-person botmaster discussed previously. That storm-warning spam message linked to a worm that installed both a downloader and a peer-to-peer client on each infected computer. The conventional, contemporary network for distributing information online is made of *client* and *server* machines—for the basic botmaster example, the infected computers in your botnet are all client machines that download the material you specify from a server, a central machine somewhere. A peer-to-peer system, by contrast, treats all of the computers on the network as *peers*, which are capable of being clients and servers simultaneously, both requesting and providing information from and to other peers. Any one of the infected computers with a complete message could route it to the others, passing data along one to the next over their diverse connections. Bots communicating among themselves as peers meant that any changes the botmaster sent out—new C&C instructions, packages of code for new functionality, spam text and address databases—could propagate out through the network with less work and traceable exposure on the part of the botmasters. The machines circulate it, one to another, on their own. The botmasters could drop new material in a few select places, like ink in a pool or a rumor in a crowd, and watch it diffuse.

Within months, this already fairly sophisticated system was broken into two networks: one managing package distribution and the other C&C, with bots passing along regularly updated directions to keep the programmers in control and the lines of communication open. Storm's authors had built a dream of decentralized and outsourced production, turning spam into the financial backer and infection vector for a global workhorse made of other people's capacity. Researchers found that Storm acted as a vast spam factory drawing on the botnet's resources. It had "a work queue model for distributing load across the botnet, a modular campaign framework, a template language for introducing per-message polymorphism, delivery feedback for target list pruning, per-bot address harvesting for acquiring new targets, and special test campaigns and email accounts used to validate that new spam templates can bypass filters."[93] In other words, the work queue kept the workload of sending spam, among other projects, evenly spread across the many thousands of infected computers, ensuring that few were underutilized. Different spam campaigns could be paced in their distribution by the botnet. In the primitive example in the previous section, the botmaster could distribute only one campaign at a time to all of his bots and would have to cancel it to start another one, whereas the Storm system could simultaneously run several different profitable campaigns alongside the all-important malware self-propagation spamming.

Individual bots could produce one unique message after another—that's the "polymorphism"—to beat filters with a tide of minor combinatorial variations, litspam text, and alternate names and subject headings.[94] The bots could report the failed messages and take the addresses, invalid or dead, off the target list of addresses to be used and add new addresses, fresh from infected machines. Evidence was found of testing systems using common third-party email services such as Hotmail and Yahoo! to fine-tune new spam campaigns and get past the basic filters. The bots on this system, given their instructions and material, each sent an average of 152 messages a minute while the notional owners of the infected computers worked on spreadsheets, answered email, played games, or left them on while out of the office. "One such [spam] campaign—focused on perpetuating the botnet itself—spewed email to around 400 million email addresses during a three-week period."[95] One campaign, it should be remembered, among many: the Storm botnet's segmentation into different subgroups of computers, with the control of each accessible by a different security key, strongly suggests that part of the business model lies in renting out capacity, piece by piece, for others to use.

Of those 152 messages a minute, only about one in six is successfully delivered, and that delivery is prior to several stages of potential filtering. The work is so inexpensive that rates of success can be far lower than even those of earlier spam systems. For instance, the address harvesting functionality of the segment of the Storm system under research analysis returned almost a million email addresses. About half of these were duplicates, and a tenth were not valid email addresses at all, with endings like .gbl, .jpg, .msn, .hitbox, and so on—a sign that the pattern-matching software looking for the characteristic email address shape (foo@bar.bat) was not very good, and many of the harvested computers contained slightly mangled addresses or things resembling addresses. So many mistakes, and so much duplication of effort, with only one in six messages even making it to the jaws of the mail filtering systems through which only some small percentage will pass: this completely unacceptable level of failures simply does not matter if the means of production and distribution are so powerful and so cheap. At 152 messages a minute from every one of many thousands of computers at no cost to you, the failure of the vast majority of messages at every stage means nothing. This is a post-scarcity manufacturing model of fantastic profligacy, recalling "The Library of Babel" as a study in Borgesian publishing economics. Somewhere in those endless hexagonal rooms of books filled with random letters is "the minutely detailed history of the future, the archangels' autobiographies . . . the true story of your death," all generated affordably if the cost of production is zero, or close enough.[96]

A new worm, taking over a new machine, will include an antimalware kit to clean its competitors off, stopping the operation of suspect files and then going through their code for likely passwords and other information to take over other computers on the competitor's botnet. The suspect programs are usually just lists of known malware files, which create a kind of found poetry of filenames with a functional banality meant to evade the interest of the user looking for malware, or to thumb its nose at them:

W32.Blaster.Worm "msblast.exe," "tftpd.exe,"
W32.Blaster.B.Worm "penis32.exe,"
W32.Blaster.C.Worm "index.exe," "root32.exe," "teekids.exe,"
W32.Blaster.D.Worm "mspatch.exe,"
W32.Blaster.E.Worm "mslaugh.exe,"

W32.Blaster.F.Worm "enbiei.exe,"
Backdoor.IRC.Cirebot "worm.exe," "lolx.exe," "dcomx.exe," "rpc.exe," "rpctest.exe"

From the struggles on individual computers to the control of global spam production, Storm did not want for rivals. It shared the upper reaches of the food chain with systems like Kraken (alias Bobax, Bobic, Cotmonger), Cutwail (which may have been responsible—again, certainty in measurement is difficult here—for about 29 percent of *all* spam between April and November of 2009), Nugache, Ozdok (alias Mega-D), Grum, Lethic, Festi, Bagle, Srizbi (alias Exchanger, Cbeplay), Conficker (alias Kido), Rustock, and Wopla.[97] This strange, small population of hundred-handed titans with evocative names is collectively responsible for the vast majority of email spam, all quickly learning from each other and fighting for market share. Their history is defined by rapidity: rapid innovation, just as rapidly copied by the others, as well as rapid increases and declines in capacity as security patches are released and the botnets steal captured machines from each other.

SURVEYING STORM: MAKING SPAM SCIENTIFIC, PART II

Among these competitors, Storm remains the best researched. As a vein of quartz suggests the possibility of gold nearby, so does spam often imply new areas of exploitation and innovation online, drawing in scientists as well as security professionals and curious hackers of all stripes. As with the problem of email corpora for scientific spam filtering, simply fashioning an epistemic object on which experiments can be performed is the difficult first step for scientists encountering the botnet. With the email corpus, the problem was one of privacy. With the botnet, it is that of the gold rush: too many teams and individuals following the same thread of quartz. The tents and campfires multiply, and every stream fills with silt. Storm is notorious in the computer security community and has some major flaws in its architecture: because every compromised computer on the network is a peer when it comes to circulating information, it can tell a lot of others where to listen for instructions, leading them astray, that is, into the labs of interested parties. These factors make it attractive to researchers who want to measure or manipulate it and to saboteurs who want to harm it.

As a botnet, Storm turned compromised computers into a platform for self-propagation, spam campaigns, and ambitious exploits, and it has in turn become a platform on which scientists, security specialists, hackers, and other interested parties launch project after project. ("It is difficult to strike a balance between being a good citizen in the [Storm] network and potentially damaging it through novel research techniques," as one group put it.)[98] Filtering out the effects of attacks and research projects being performed on the botnet is one of the hardest parts of doing research on Storm. Like the wonderful scene in G. K. Chesterton's metaphysical detective novel *The Man Who Was Thursday* when the anarchist conspirators realize that they are all secret police agents attempting to infiltrate the anarchist conspiracy, Storm researchers keep encountering other researchers and the results of their work in the botnet itself.[99]

A surprising flaw in the Storm system—a bad pseudorandom number generator that produced a recognizable pattern of IDs that were internal to the Storm network itself, rather than the outsiders exploring and traversing it—made it possible for scientists to gradually separate out and define a population of other users. This cohort is a population of buggy and broken bots, "vigilante researchers, rival spam gangs," and other players, all seeking to slow the system down, test it out, and make it impossible for the Storm bots to communicate with the Storm botmasters or interfere with the other onlookers. Rather than the kind of monolithic artificial intelligence dominating the network as imagined by science fiction, such as Wintermute in William Gibson's *Neuromancer*, as a total and enclosed apparatus—"Case laughed. 'Where's that get you?' 'Nowhere,'" the AI replies, "'Everywhere. I'm the sum total of the works, the whole show'"—we find instead something more like a gold rush boomtown or an Arctic research base, criss-crossed by natives and scientists, crooks and surveyors looking for a cut, sociologists, cops, and broken machines: a gathering place for interested parties.[100] Sometimes the gold is gone but the town remains: "There was a joke at a recent security conference that eventually the Storm network would shrink to a handful of real bots and there would still be an army of rabid researchers fighting with each other to measure whatever was left!"[101]

Of the population of visitors and immigrants to this outpost, built on flows of spam as other communities were made on flows of railroad tracks or grant money, security groups and the agents of the government and the military have become some of the most prominent. "The more worrying thing is bandwidth," said a security analyst of Storm at its likely peak (its peak and its total size being objects of considerable debate). "Just calculate four million times a standard [high-speed Internet connection]. That's a lot of bandwidth. It's quite worrying. Having resources like that at their disposal—distributed around the world with a high presence and in a lot of countries—means they can deliver very effective distributed attacks against hosts."[102] Storm, or its owners, seemed to periodically identify attempts on the part of serious security firms to investigate it and would retaliate with DDoS attacks, like Mydoom's swamping of SCO, Inc., with requests from its bot computers. Sometimes they could take an investigator's servers down for days. "As you try to investigate [Storm]," said Josh Corman, the host-protection architect for IBM's Internet Security Services, "it knows, and it punishes. It fights back."[103]

The question of jurisdiction raised by these attacks is a very real one: at their largest scale, at the size of Storm and Wopla and Srizbi and Cutwail, botnets have strange relationships with national boundaries and human populations. Bot computers make the botnet grow by sending out self-propagation spam (as well as using more esoteric means like those files Mydoom seeded in file-sharing applications for others to discover). To spread the botnet infections, by spam or other means, the compromised computers need to be on, and online—an obvious fact with a strange implication: spam can be seen to rise and fall, and botnet propagation spike and diminish, as the earth rotates. The terminator, the line that separates day from night, is part of the circadian clock of large botnets, a diurnal rise and fall in total capacity and rates of potential infection.[104] The infrastructure of the botnet apparatus also changes, more slowly, with shifts in global Internet access. The next great botnet resource, many agree, is the African continent, home to about 100 million PCs, of which an estimated 80 percent are compromised or infected with some kind of malware. Most of the boxes are running pirated operating systems (and therefore may not receive security updates and patches) and their owners can't afford antivirus software (a standard Windows installation license can be laughably expensive relative to local salaries), both of which make them significantly more vulnerable. Most of the Internet access has been telephonic dialup—which is to say, fairly useless for a botmaster—but a great push to connect the continent to the big cables that form the global backbone will bring in a huge population of additional, accidental victims for the cloud.[105]

Finally, the success of spam and self-propagation messages, as well as many particular aspects of the exploits that worms perform, depend on languages. That malware download pitch promising a news story would not get much traction with a user who reads only Mandarin Chinese, Russian, or Hindi, and the installation process by which the worm takes control might rely on code in the language-specific version of an operating system. Different botnets therefore have different demographic dynamics: the perspective of the botnet sees national boundaries as relevant only insofar as different economies and infrastructures affect the number of computers online. Botnets operate in vast regions whose edges are language, software, and time zone rather than borders. The jurisdictional issues are beyond complicated. A botnet apparatus, setting aside the global population of infected computers, might be using many hosting services under many identities in many different countries, all hooked up into an interdependent system. It is here, at the farthest perspective and the broadest spatial scale, where the borderlines have almost evaporated and Pitcairn Island is one minute node among many in the botnet's architecture, that the transition from tool to weapon appears: the boundaries are violently reasserted, and nations and their armies get into the business of spam and its consequences.

THE OVERLOAD: MILITARIZING SPAM

"Hackers often use botnets to generate spam, but their real strength lies in their ability to generate massive amounts of Internet traffic and direct it against a small number of targets," explains Col. Charles W. Williamson III in his profoundly strange article "Carpet bombing in cyberspace."[106] Email spam, despite being instrumental to the propagation and finances of botnets, has stopped being their reason for existence—despite the fact that they now produce almost all of it. In many ways, it has become the most boring part of their operation: 120 billion messages a day surging in a gray tide of text around the world, trickling through the filters, as dull as smog. It is still what they do, but the technical excitement is elsewhere now, as are the fascination and the panic. It is in the prospect of DDoS attacks—massive exploits that can temporarily kill the networks of companies and countries—and in enormous amounts of computing power available for cracking codes and finding passwords, as well as in a new market for accidental intelligence data. Spamming, though remaining largely unchanged, has become a minor and incidental part of the system——a technique disappearing into ubiquity rather than obsolescence, having been reinvented as part of a new language of threat.

In April 2007, the Estonian government provoked an international incident by removing a bronze statue of a Soviet soldier from the center of Tallinn. That statue was an object with several voices: the product of multiple histories and an intersection of numerous timelines, identities, and archives. Like the *Enola Gay*, it meant very different things to different people, and the delegations and cohorts gathered there did not agree at all. For both Russian nationals and the ethnic Russians in Estonia (who constitute a quarter of the country's population), the 1947 statue—which was (or was not) erected over the graves of fourteen (depending on who counts) Soviet soldiers—stands for those who fell in the fight against Nazi Germany. For Estonians, it symbolizes the soldiers who took Estonia back from the Nazis and then did not leave, occupying the country until 1991. The statue had become a point of continuous friction between the ethnic Russian minority and Estonian nationalists and police, and the removal of the statue took place in a scrum of rioting, rubber bullets, hurled stones and bottles, and television coverage.

Almost immediately thereafter, Estonia's network traffic started to surge. The servers for several major Estonian institutions, including government ministries, banks, and newspapers, were hit with massive spikes in activity, enough to eat up their bandwidth and repeatedly take them offline. The Estonian newspaper *Postimees* was swamped with comment spam and millions of page requests from countries such as Egypt, Vietnam, and Peru (that is, countries unlikely to have a major interest in Estonian affairs). Official government sites were hacked and redecorated with anti-Estonian visuals and rhetoric or simply driven offline by repeated bursts of traffic. Many of the country's official organs were be unable to get the word out about events within their borders, and in a country as small and Internet-driven as Estonia, where 90 percent of the banking transactions are handled online, the loss of official web services was invasive and distressing. By mid-May, the Active Threat Level Analysis System (ATLAS), a data gathering tool run by the security firm Arbor Networks, provided a partial picture of the events: 128 distinct DDoS attacks over two weeks against a handful of crucial Estonian sites. "Someone is very, very deliberate in putting the hurt on Estonia."[107]

The Estonian DDoS attacks provide a deeply unsettling perspective on the vulnerabilities of web services, particularly for small countries—and Estonia is thoroughly wired, and one of the countries most reliant on Internet connectivity for the daily lives of its citizens—as well as on the state of post-Soviet diplomatic relationships and the new forms of subwar harassment that countries can exert on each other. Those facts aside, this string of events immediately became an argument in which botnets and spam were fashioned into objects of geopolitical, military concern and "cyberwar" hype: they became attention-grabbing sources of rhetoric.

Where there is one malware infection, there is almost always more than one, and conflict and competition between them, and so it seems to be with the narratives told by different constituencies during technological dramas. Where there's one story, there are many, and they do not fit conveniently together. From the network security perspective, the DDoS attacks, related exploits, and floods of spam against Estonian sites were a serious matter, particularly for a small country with relatively low bandwidth capacity to absorb them, but they were also entirely familiar and could be handled with technical aplomb after the first rush of panic.[108] Estonian and international security services could track the traffic, block the clusters of Internet addresses responsible for most of it, work closely with service providers, and engage in other defensive measures to mitigate the effects. Rapid response and knowledgeable security managers and system administrators, in the case of Estonia—as in many similar attacks on diverse companies and countries—could undercut a sustained attack.

From the perspective of official governmental statements, though, the April and May attacks were a very different matter, constituting a "cyberwar" attack or one example among several of a "digital Pearl Harbor."[109] Or a digital Hiroshima: "When I look at a nuclear explosion and the explosion that happened in our country in May, I see the same thing," said Ene Ergma, president of the Riigikogu (Estonian Parliament). We are abruptly in a whole different class of metaphors: "Like nuclear radiation," she continues, "cyberwar doesn't make you bleed, but it can destroy everything."[110] This obscene analogy, comparing a series of infrastructural slowdowns and panics produced by DDoS attacks with the mass death and devastation of a nuclear weapon, epitomizes the process by which spam, and its technical transition into the botnet, was adopted into political and military narratives. Estonia is a NATO country, and there was consideration of invoking Article 5, which mobilizes all NATO members against an aggressor who has attacked one of the member countries, thereby initiating the first war in which spam played a major role. The Estonian attack precipitated the creation of a NATO Cooperative Cyber Defense Center of Excellence in Tallinn. During the attacks, "NATO dispatched two observers to Estonia and the Americans sent another in order to 'observe the onslaught.'"[111] Col. Williamson, advocating the construction of a U.S. military botnet, asks: "Can the U.S. reasonably believe that other nations have not learned from the DDOS attacks on . . . Estonia in 2007?"[112] We have seen the processes scientists go through to turn email spam and botnets into the kinds of epistemic objects they need for their research; in these remarks, we can see the process by which the botnet becomes a militarized object—a matter available for strategic analysis, countermeasures, and deterrence.

In the military language of botnets at war, spam is a sinister process of mobilization, as infections spread and botnet capacity is built. Even as that rhetorical turn is underway, however, the place of spam in the public perception of the network has changed. Complaint and survey data in the United Kingdom and the United States suggests that after the millennium, even as spam was beginning one in a series of massive growth spurts, users became more tolerant of spam as an insignificant matter, increasingly regarding it more as a nuisance and less as a threat: just something to be filtered, coped with, deleted, and ignored.[113] From the perspective of the vast statistical majority of users, spam does not even really seem like a crime, much less a cybercrime. We expect cybercrime to be big, dramatic, and exciting—the prosecutor keeping hacker Kevin Mitnick in solitary because (so the cybercrime fantasy went) with a moment's access to a telephone he could whistle the secret launch tones that would start a nuclear war[114]—not the quotidian trickle of fake bank notices, hilariously maladroit scams, and ads for porn and pills. And it is a trickle, for many, with the apogee of sophisticated techniques applied to big data by service providers creating truly effective filtering systems such as those Gmail uses. (As I was writing this book, I often got into conversations with people who would mention that from their perspective, spam seemed to have "gone away," by and large, or become negligible as a part of daily life.) Spam was starting to seem more like an irritant, a kind of mild chronic problem that had ceased to be of much significance and become an operational inevitability, a cost of doing business for the individual user—and a business in itself for the security provider.

The alliance of spam and malware that produced the botnet architecture also produced a new business for security professionals. "'Antispam is a big business now,'" Jessica Johnston quotes a researcher as saying, "'something that the large corporate customers are prepared to pay for . . . The early antispam products were always free or relatively cheap.'"[115] In those same early days, spam was still seen as a social problem and possibly a legitimate marketing opportunity. Now it has been recast as a far more consequential and problematic object, wedded to the enormous exploit-enabling machinery of the botnet, a matter of concern for the big-ticket culture of enterprise security firms. (Airports from Pudong to JFK have ads for products like Barracuda Networks' "Spam and Virus Firewall"—"Blocks e-mail borne spam and virus intrusions while preventing data loss"—to attract the eyes of business travelers.) It has also become an area of interest for the much bigger-ticket world of the military, just when the civilians were getting used to it and starting to see it as a part of everyday life.

Threat or annoyance, spam in the shadow of the botnet is repeatedly rescripted by enterprise security groups and the military. "'If the fraudsters destroy e-commerce as we know it . . . it's going to do us a lot of harm,'" says another researcher. "'If the fraudsters undermine the banking system, and there is every indication that they're close to doing that through insecure mirrors and proxies all over the net so you can't see where it's coming from, then in all honesty, that does far more harm than knocking down a couple of towers and the like. No lives are lost, but even so, the overall impact is greater.'"[116] The comparison to September 11 recalls Ene Ergma's exercise in atomic metaphor after the Estonian DDoS attacks. The point is not to question the premise for this comparison—whether the financial damage caused by spam, and the potential cost of a loss of confidence in online banking and ecommerce, exceeds the financial impact of 9/11—but to observe these metaphors in operation.

Even as the number of users who could remember a network on which spam was still something new and startling steadily declines relative to those who have known nothing else, big institutions give it a fresh coat of paint as a threat of very grave consequence. Antispam is no longer the area of the communal hobbyists, activists, and vigilantes gathered on NANAE, or the collective of programmers building better Bayesian filters. It's now part of Homeland Security, a front in the "cyberwar," a place for private contractors to overlap with officers from the Air Force Cyber Command, NATO, and the FBI. The DDoS has also made a strange lateral move into protest events, becoming the weapon of choice for online activist groups such as Anonymous. Programs including the grandly named "Low Orbit Ion Cannon" (from a superweapon in the science fiction game *Command & Conquer*) enable individuals who download it to voluntarily join a botnet. This public-spirited botnet can then be directed to attack sites like those of organizations that were hostile to WikiLeaks and of repressive governments like Syria's. The values of these technologies, and the narratives in which they can be enlisted, are in constant transformation.

CRIMINAL INFRASTRUCTURE

Though the botnets rely on distributed computers, the business of email spamming has become far more centralized. The economies of scale that make spam possible demand volumes of messages that only a major, sophisticated, evasive, and inexpensive infrastructure such as a botnet can provide. The days when hundreds of dubious bit players with some office space, a couple of rented high-bandwidth connections, and a bunch of cheap PCs with off-the-shelf mail marketing software could build a business around stock touting and potency pills are long past. The combination of filters, responsible service providers, legislation, and informed consumers have swept out those small-timers with their pill-financed convertibles, entrepreneurial zeal (recall those offers in Rodona Garst's IMs: "I now have that mortgage deal, cable boxes, anabolic steroids and Adult If they want to") and phones ringing with the charivari's threatening calls. Those left are the cohort, the few hundred groups responsible for more than 80 percent of spam, who have the training and the capacity to leverage the network to generate the hundred billion–plus messages that constitute the daily spam load. Even as their systems spread to encompass the globe and traffic in numbers and amounts difficult to grasp, the group at the core of spam shrinks steadily into one aggressive and bickering extended professional family.

Similarly, the infrastructure that enables their activities has become more centralized. "Our datacenter is situated in top-level modern MarketPost-Tower IT center, San Jose, CA, USA": so runs the text on a defunct website site belonging to McColo Hosting Solutions. (The site's bare-bones text lives on at the Internet Archive.)[117] McColo had a reputation among the loose confederation of private- and public-sector security professionals, IT analysts, and cops as a "bulletproof hosting" provider—a term that goes back to the early days of NANAE, referring to an ISP that will not kick clients off regardless of the complaints that they receive and that is thus a haven for spammers. If you paid extra, they would take the flak of complaint and criticism for your activities and even take steps to disguise your existence—allegedly doing this by moving some of their offending clients to different subnets, like publicly firing a problematic employee in one department and quietly hiring them into another. McColo was hosting the servers for the C&C channels, many of the web pages for moving products and malware downloads (rxclub.biz, high-quality-viagra.com, pills24.biz, valium-plus.com, etc.) as well as anonymization and proxy services and payment sites for several major botnets, including Srizbi, Mega-D, Rustock, and Cutwail, along with some other nefarious content. On the November 11, 2008, the two "upstream" providers for McColo—the companies whose backbone Internet connectivity McColo relied on to run its hosting service—cut off their bandwidth after receiving reports on their activities. Global spam activity abruptly and precipitately began to drop by the millions and then billions of messages. At the lowest point, global spam levels declined by roughly 65 percent.

The forces involved in the shutdown of McColo included journalists, security analysts, and the administrators of the major hubs that provided McColo's connectivity. (Its shutdown left a strange dead zone in the Internet's address space: the block of addresses allocated to McColo had ended up on enough blacklists for their bad activity to render others leery of taking them over, leaving them as "ghost number blocks," like a house known for its suicides and shunned by potential tenants.) We can see a similarly mixed population in the Mariposa Working Group, which came together to shut down the Mariposa botnet: an international collection of security specialists working with the FBI and Spain's Guardia Civil. If the very concept of Internet governance is presently diffuse, so is its enforcement, with loose working groups that overlap jurisdictions and expertise, odd bedfellows in some cases—like the Finnish security specialists, NATO and U.S. observers, and Estonian ISPs brought together by the DDoS attacks on Estonia in 2007—that form in relation to the diffusion of the problem.

Though we seem to have come a very long way from Peter Bos's message of conscience to the terminals supported by MIT, this history can also be read as a kind of interregnum, a transit from one period of overt control by systems administrators to another. The sysadmins of the early years of the network, Gandalfian figures maintaining order in their domains according to their lights, have become what Alan Liu terms "a priesthood of backend and middleware coders" as well as a small expert elite of security analysts, state agents, and ISPs.[118] Users can take refuge within the relatively spam-free zones that the developers build, such as Gmail and Facebook, with robust filtering and community management, paying with advertising and their personal information and user activity—with their quantifiable attention.

Imagine another industry that could drop in production by more than half overnight with a single industrial action or largely vanish if a few hundred people were imprisoned. Conventional email spamming has long since passed a peak of easy money and is well into the hard grind of optimization and efficiency, trying to extract the maximum value from the network in a dense matrix of constraints. Spam levels rebounded over the weeks after the McColo shutdown, as the botmasters found new ISPs willing to work with them and host their systems and moved the bots over to the new command channels, but the revelation of just how small the industry had become was clear. The conventional spam with a heritage running alongside that of email and the rise of the web had become the world's most efficiently concentrated business. Spam's history of labor-saving solutions, like Canter and Siegel's Usenet-spamming script and the early pattern-seeking address harvesters looking for "xx@xx.xx," which leveraged the automated accumulation of many small effects over a vast public infrastructure, has made it possible for group of people about the size of a very small town to affect part of the daily lives of the planet's entire computer-using population.

Our history began with networking computers together and then connecting up the networks for the sake of efficiency and resource sharing and remote access. Our story ends with a small group of criminal spammers with remarkable talent and vision, stitching networks of malware-infected personal computers around the world together into globally distributed machines devoted to sending spam—and to other, more sinister tasks—for efficiency, resource sharing, and remote access. This is a chapter in the shadow history of the Internet. It is in some ways akin to the obverses of globalization, with the construction of covert markets, franchised criminal organizations, and massive supply-and-demand logistics for operations such as drug smuggling, counterfeiting, and human trafficking that parallel, parasitize, or undergird those of conventional globalized operations.[119] Cloud computing is an immensely popular model in contemporary business: order up some given amount of computing power from Amazon's services or a company such as Rackspace, set up an instance of an operating system, and from your laptop control processing power and bandwidth that would have been inconceivable to any backbone-administering baron of the Usenet years—with the cloud computing provider handling the software maintenance and the security of the server racks being cooled in an anonymous facility somewhere. However, if you run a different kind of business, you could set up a deal with the Conficker botnet at its peak, with access to millions of computer systems distributed across 230 top-level domain names (that is, scattered across many countries and hosts), order the amount of bandwidth you need and the appropriate operating system, and start running spam campaigns, DDoS attacks, data harvesting, or password cracking, as you wish.[120] (Security researcher Robert Hansen has made the point that this activity can change the dynamics of corporate and state espionage: don't start with trying to infiltrate a company or a government, but instead give a botmaster a list of the Internet addresses or machines you're interested in, and if they've already got them on the network, you can simply buy in directly, and start exfiltrating information.)[121]

This is one form of the end of spam: its subsumption into criminal practices and systems of far greater power, profit, and complexity—indeed, nothing less than the construction of a *criminal infrastructure*—as a mere source of funding, that is, one of a suite of services and part of standard operating procedure. It also implies the possibility of another end of spam, though, one that many people desire and that events like the McColo shutdown emphasize. The stitched-together networks of machines that constitute the botnet, the whole spam apparatus, are unexpectedly fragile. They reflect the same tension between distributed and centralized that plagues cloud computing as a whole. A few pivotal arrests—or threats of arrests, as in the case of Rustock botnet, whose controllers apparently abandoned it under increasing legal pressure—and the volume of email spam plummets by billions of messages per hour, though it quickly climbs back up as newcomers enter the business.[122] A remarkably small number of registrars handle the bulk of the registration of domain names for spam sites; the same is true for hosting and other Internet services.[123] Though it can be relatively easy for groups as sophisticated as contemporary botnet administrators to switch providers, it is still time-consuming—and the production of a climate of reasonable fear among service providers who might otherwise be tempted to take some spam business could make the work of migration much more difficult.

The financial infrastructure behind the consolidated botnets is similarly brittle. One group of researchers found that 95 percent of "spam-advertised pharmaceutical, replica, and software products are monetized using merchant services from just a handful of banks."[124] A "handful," in this case, meaning three: a Norwegian-owned bank in Latvia called DnB Nord, the Azerbaijani bank Azerigazbank, and the St. Kitts-Nevis-Anguilla National Bank in the Federation of Saint Kitts and Nevis, in the West Indies. (Since the start of this research into the banking system, spammers—nothing if not adaptable—have migrated away from Azerigazbank to two other Azerbaijani institutions.) As the researchers point out, finding payment processors willing to do business with spammers is not a trivial matter, and there aren't that many of them. They propose a powerful demonetization strategy: a swiftly updated financial blacklist of institutions for which Western banks will refuse to settle a small subset of transactions. The money in spam, aside from phishing, 419-type scams, and businesses spun off from excess botnet capacity, comes from Westerners paying with their credit cards online for a very narrow range of products (pills, fake watches, cracked software, and the like). If you can bracket out card-not-present transactions for that set of products, identifiable by their Merchant Category Codes, to that small collection of banks, you could essentially halt the circulation of a large portion of the funds that keep email spam in business. Asking issuing banks in the United States not to honor certain transactions may seem a radical step, but it has been done before in relation to some online gambling transactions.[125] (Though the chaotic record of that gambling-regulation project suggests the many layers of law, policy, jurisdiction, and enforcement that such a spam-halting project would confront: struggles with Antigua and disputes with the World Trade Organization (WTO) over trade agreements and "secret" trade settlement concessions, the proposal of alternate bills, and the indictment of online poker sites for colluding with payment processors to disguise gambling transactions as innocuous purchases of golf gear and jewelry. It would not be simple.)[126] Just as with Rodona Garst and her team, who hopped from one hosting provider to another, spammers are critically dependent on the availability of infrastructural access, which is why they began to build their own. The points of failure for their operation lie there.

A few carefully directed and executed interventions could make an enormous dent in the production of email spam. Filtering and laws did not stop it, by any means, but they have painted it into a developmental corner with severe bottlenecks: an almost totally centralized, consolidated business dependent on colossal volumes of mail to survive. Even assuming such an intervention were to be successful, however, that event would stop only one of the forms that spamming has taken—admittedly, one of the most visible and hardiest. Recall the proliferation of search engine spam and the flood of spam comments in blogs; wiki spam; the subculture of Twitter spambots piggybacking on popular phrases with their untrustworthy links concealed with address-shortening technology; social network spamming inside services, such as "likejacking" in Facebook; line-blurring cases like content farms, link-baiting blog posts and sleazier forms of attention-grabbing viral media: these will not go away, and only a few offer the same obvious (if politically complex) points of failure that email spamming does.[127] (And this list does not even mention forms of spam now being born, including "spam books" and spam in online games.) Spam persists and diversifies because we are living through a major, complex transition in the constitution and management of our own attention, a transition moving faster than our governance, our metaphors, and our software can keep up with. Spammers—the disbarred lawyers, impoverished con artists, would-be pornographers, credit card thieves, and malware coders—are the avant-garde, the wildcatting exploiters of this transition. They find domains where salience is being generated, whether in a comment thread, a search engine result, a social media platform, or your email inbox, and move to commandeer it. They are the crudest and most abject form of this capture, from students pranking each other with the words of a *Monty Python* sketch to global botnets producing more email than everyone else on earth, every single day. In their crude way, they show the rest of the online population the network's new capabilities, the new forms of attention and community experience, which we have not yet fully understood.

CONCLUSION

What, then, is spam?

"Spam" is a word with remarkable properties. As this history has demonstrated, the meaning of this word has remained surprisingly consistent through enormous shifts in technology and scale. At the very outset of this book, in the fourth paragraph, I emphasized the diversity of values assigned to "spam." Doing so was necessary to frame the argument and the history for readers who thought of spam solely as unsolicited email for timeshares and Cialis and to introduce them to a term that encompasses time-wasting loquaciousness on Usenet in the 1980s and the planet-size criminal infrastructure of botnets in 2010. It is a single word applied contemporaneously to projects with completely distinct technical means and social motives and retroactively used to dub behavior on antediluvian systems whose properties would be virtually unrecognizable to someone casually retweeting an Instagram image from their phone today. What this history makes clear is that a consistent, continuous, intuitive, and precise value underlies this menagerie of applications.

What is "spam"? *Spam is the use of information technology infrastructure to exploit existing aggregations of human attention.* That is the meaning of "spam" once all the technological particulars of search engine spamming or phishing campaigns have been worn away. In the process of breaking down this meaning, we can also summarize what spam demonstrates and what it can indirectly teach us.

THE USE OF INFORMATION TECHNOLOGY INFRASTRUCTURE . . .

First, spam is an information technology phenomenon. Across their many modes and domains, spammers push the properties of information technology to their extremes: the capacity for automation, algorithmic manipulation, and scripting; the leveraging of network effects and vast economies of scale; distributed connectivity and free or very low-cost participation. Indeed, from a certain perverse perspective—one with consequences for my argument—spam can be presented as the Internet's infrastructure used maximally and most efficiently, for a certain value of "use." Consider email spam—all those millions of messages cranked out by thousands of computers around the world in hopes that a vanishingly small amount will get through to the eyes of that fraction of a percentage of people who will actually respond to such messages (and get their credit card information stolen). Spammers will fill every available channel to capacity, use every exploitable resource: all the squandered central processing unit cycles as a computer sits on a desk while its owner is at lunch, or toiling over some Word document, can now be put to use sending polymorphic spam messages—hundreds a minute and each one unique. So many neglected blogs and wikis and other social spaces: automatic bot-posted spam comments, one after another, will fill the limits of their server space, like barnacles and zebra mussels growing on an abandoned ship until their weight sinks it. Under DDoS attacks, servers do what they're built for, to the maximum of their capacity: they serve web pages so rapidly and in such quantity that they can no longer provide them to anyone else.

This description is obviously ludicrous, a panegyric of pure function, while still being true. Although some of spam's forms have a surface resemblance to junk mail, junk faxes, and telemarketing calls, it is native to networked computers and draws its peculiar properties from those of the technology it uses. Two consequences spring from this fact, both significant in the struggle to stop spam. One is the failure of previous metaphors to accurately describe what spam is like for purposes of specifying laws or creating programs. The other cuts deeper, and is implicit in that little parody of the technological sublime earlier, with spam as the hyperthyroidic, maxed-out version of existing uses of the Internet. Spammers take advantage of existing infrastructure in ways that make it difficult to extirpate them without making changes for which we would pay a high price, whether in the hobbling of our technologies or in contradicting the values that informed their design.

We could have a network almost wholly without spam of any sort, torn out root and branch, if we were willing to live with the consequences. It would be a carefully specified theme park of a system, another civic-minded network like France's Minitel or a highly managed proprietary space like AOL, whose purpose and goals were clear and whose users could be closely controlled. (In a more current example, it might resemble the aggressively vetted Apple App Store.) Although, as has been discussed, even Minitel was turned to unintended purposes, and AOL's iconic walled garden for Internet amateurs and easy marks for spammers was hacked by its own population of bored teenagers and turned into a file-sharing and prank-pulling machine.[1] It remains difficult and perhaps impossible to truly out-think and out-design all the ways in which a technology will be reinvented, to guess all the counterstatements made in the technological drama. Nonetheless, we could reliably make spam in general far more difficult to practice by changing the infrastructure, but the value lost would be higher than the value gained. We prize the speed at which we innovate, the volume of data to which we can get access, the openness and ease of our communications channels and social interactions, and the utility of having anonymity and ambiguity available in our technical specs and our social mores.[2]

. . . TO EXPLOIT EXISTING AGGREGATIONS OF HUMAN ATTENTION

The second element of the consistent meaning of "spam" builds on the first: "to exploit existing aggregations of human attention." Spammers find places where the open and exploratory infrastructure of the network hosts gatherings of humans, however indirectly, and where their attention is pooled. The use they make of this attention is exploitative not because they extract some value from it but because in doing so they devalue it for everyone else—that is, in plain language, they waste our time for their benefit. (Recall the Canter and Siegel spam campaign on Usenet, the objection against which wasn't simply that they were acting commercially but that they didn't respect salience, barraging everyone indiscriminately with their lame message, treating the whole network as a passive audience whose time was theirs to spend.) Again, two consequences follow from studying spam in this light: we can see the history of networked computing as a thread in the history of the management and distribution of attention, and we can see in concrete terms how the nebulous shape of community online used spam to define itself and how it evolved.

Seen in terms of attention, the use of the word "spam" to describe web pages meant to manipulate search engine results doesn't seem as strange as it might when the word first made its way from people dominating chat channels with their yammering. Of course, we are paying attention when we're participating in a chat discussion or a Usenet thread, and we pay attention to our email inbox, turning the space of email addresses into a vast reservoir of awareness—but, in a somewhat more abstract sense, search engines are ways of consolidating indirect attention. A system such as Google's turns the work of individuals making pages and linking to things into a reservoir of votes and decisions—this website is better than that one if you're interested in this topic—that is refined and presented in response to a question. This system is indirect in that rather than querying people directly and demanding their notice, it produces these answers as a side effect of something people were already doing for themselves: linking across the network and making connections.

This system produces a strange, distinctively contemporary kind of group, formed without the intention of its members, whose machine-aided human work is aggregated and adapted by machines and algorithms into a coherent product for human eyes. Unless you very deliberately use code that forbids Google's spiders from indexing the words and links on your pages, you are folded, quietly and without comment, into a very rarified form of machine-managed community. It is not an imagined community, but an unimagined, amorphous one, constantly reconfiguring according to rules and protocols. We can call it the "coordinate coexistence" (from Yochai Benkler's term) in which we make meaning and share knowledge and resources without necessarily making deliberate choices to cooperate and participate.[3] Google's search box is the narrow channel through which rushes a torrent of human noticing, a constant source of attention as much as any inbox, Twitter feed or Facebook wall, and therefore ripe for capture. Attention is the commonality between the different platforms, the constantly scarce resource even as the price of memory plunges and available bandwidth capacity multiplies and multiplies. Spammers help us see how it is collected and transacted by showing up in the most unlikely places to flood the channel and exhaust the goodwill of those who have directly or indirectly assembled there.

In exposing-by-exploiting the flows of attention on the network, spammers also trace out the shape of communities online. From the earliest models of the polylogue and the supercommunity at the outset of this book, through the barons and democrats of Usenet, the activist teams of charivari and NANAE antispammers, the users of social networks, and the abstract communities of bloggers and Google searchers, spam has always been there—as threat, as concept, or as fact. It provokes and demands the invention of governance and acts of self-definition on the part of those with whom it interferes. They must be clear about misbehavior, bad speech, and abuse of the system, which entails beginning to talk about what is worthwhile, good, and appropriate and, more broadly, what the point is of all this conversation, gathering, and shared time. Without having spam in the picture, we do not entirely understand much of the foundational conversations that continue to shape life together online.

In this regard, it will be particularly interesting to follow spam into the future. Already, new forms are appearing around new media. As this book was being written, Twitter spam took off, as did the first intimations of a new genre of machine-generated "spam books," colonizing the new environment of booming ebook and print-on-demand book sales with cryptic and haplessly bizarre documents that can do a brisk business, by spam standards, on Amazon.[4] Shadowing novel forms of spam as they develop leads us to new or changing aggregations of attention and forms of community, whether these are labor-coordinating engines like Amazon Mechanical Turk, gamer populations in online role-playing worlds, or the anonymous and ephemeral wells of esoteric humor that create so much contemporary Internet culture. Documenting spam over the course of this book has produced an obverse portrait, the shadow history of the Internet, and keeping an eye on its dynamics now will lead us to the strange zones of infrastructural adaptation to come.

A final observation comes from the history this book has covered, which the previous definition summarizes. We have seen spam made by humans and nonhumans, for commercial and noncommercial purposes, on networks where every character counted and on networks of effectively unlimited bandwidth. What remains—what is common throughout—is a disregard for the time and attention of others. This concept suggests a converse that's worth thinking about. If "spamming" at the most general level is a verb for wasting other people's time online, can we imagine a contrary verb? That is, can we build media platforms that respect our attention and the finite span of our lives expended at the screen? How would all the things transacted on a computer screen look if they took our time—this existential resource of waking, living hours in a fragile body—as seriously as they could? A careful arrangement of meaningful information relative to our unique interests, needs, and context. A graceful interjection at the right time, a screen that does not demand a look but waits for a glance, words that are considerate that we humans and not our filters will be reading them. It could be anonymous or rude and obnoxious, produced by machines and algorithms or by humans or crowds of humans, but it reflects respect for the attention of its recipient. It might work something like a book—indeed, like this book, I hope, which is still, which does not distract, which waits for you to refer to it, and which you can now close.

NOTES

INTRODUCTION

1. Hawkesworth, *An Account of the Voyages Undertaken*, 277–278.

2. Sophos Ltd., "Pitcairn Islands Relays Most Spam Per Person, Reveals Sophos."

3. Paasonen, "Irregular Fantasies, Anomalous Uses," 170.

4. Analyses of attention in our current technological system are themselves in a state of proliferation. For further consideration of this topic, at greater length than we can do here, see Hayles, "Hyper and Deep Attention," 187–199. There is extensive commentary on Hayles's concepts, and the addition of the idea of "hypersolicitation"—that is, the increasing sophistication of demands for attention—in Stiegler, "Taking Care of Youth and the Generations," 72–83. See also Terranova, "The Bios of Attention," and Kwinter, "New Babylons." The framework for thinking about the role of "style" in an information-rich culture can be found in Lanham, *The Economics of Attention*. Finally, there is the classic: Simon, "Designing Organizations for an Information-Rich World."

5. Pfaffenberger, "Technological Dramas," 282–312.

6. Wohl, *The Spectacle of Flight* and *A Passion for Wings*. See also Lindqvist, *A History of Bombing*.

7. Le Corbusier, *Aircraft*.

8. Scott, *Seeing Like a State*.

9. The phrase "government machine" is taken from Jon Agar's *The Government Machine: A Revolutionary History of the Computer.* Agar makes the striking point that the historical development of computation in governance was partially a process of the capture of powerful metaphors. If a government is a machine, or includes many mechanical elements, then it demands efficient mechanization—the province of the "expert movements" who seize institutional power with technological change.

10. For this transition, see—among others—Akera, *Calculating a Natural World*; Turner, *From Counterculture to Cyberculture*; and Markoff, *What the Dormouse Said.*

11. Kelty, *Two Bits*, 113.

12. Moore, "Epilogue," 234–235.

1 READY FOR NEXT MESSAGE

1. What follows here is necessarily only a brief overview of a fascinating chapter in the history of the twentieth century. A few exemplary studies that informed this short contextual section include: Abbate, *Inventing the Internet*; Hafner and Lyon, *Where Wizards Stay Up Late*; Ceruzzi, *A History of Modern Computing*; Ryan, *A History of the Internet and the Digital Future*; Markoff, *What the Dormouse Said*; Kelty, *Two Bits*; Levy, *Hackers: Heroes of the Computer Revolution*; and the fascinating analysis of ARPANET and the substance of archives in Gitelman, *Always Already New*, 97–121.

2. Levy, *Hackers*, 147. See also Szpakowski, "Community Memory."

3. Wu, *The Master Switch*, 45–47.

4. Scott, "Episode 4."

5. Feenberg, *Alternative Modernity*, 144–166.

6. Kelty, *Two Bits,* 166–177.

7. Abbate, *Inventing the Internet*, 115.

8. Hafner and Lyon, *Where Wizards Stay Up Late*, 12.

9. Taylor, interview by Aspray, 34.

10. Naughton, *A Brief History of the Future*, 75.

11. Abbate, *Inventing the Internet*, 46.

12. For more on Licklider's remarkable career, see Waldrop, *The Dream Machine.*

13. Licklider and Taylor, "The Computer as a Communication Device," 32.

14. Ritchie, "The Evolution of the Unix Time-sharing System," 1577–1593.

15. Van Vleck, "The Who Command."

16. Hafner, *The Well*, 42.

17. Dewey, *The Public and Its Problems*, 184.

18. Kendall, "Community and the Internet," 309–325, 309. Williams, *Keywords*, 76. For an overview of the complex and contradictory ideas around "community,"

especially as it enters the virtual, see Cavanagh, *Sociology in the Age of the Internet*, 102–119.

19. Rheingold, *The Virtual Community*, 5.

20. Rheingold, *The Virtual Community*, 64.

21. Coate, "Cyberspace Innkeeping."

22. Galloway, "Position Paper."

23. Nissenbaum, "Privacy as Contextual Integrity," 119–158.

24. Kelty, *Two Bits*, 3, 30.

25. Dewey, *The Public and Its Problems*, 126, 141.

26. Bygrave and Bing, *Internet Governance*, 50. Pfaffenberger, "'If I Want It, It's Okay,'" 384.

27. Joy, interview by Kim, "The Joy of Unix."

28. Jussi Parikka and Tony Sampson have written a delightful analysis of the sketch itself, pointing out that its humor, and indeed much of the *Monty Python* troupe's humor, is built around a communications breakdown, in this case the limit test of one of Shannon's channels—finding the point where noise on the line overwhelms any particular signal ("On Anomalous Objects of Digital Culture: An Introduction").

29. Parry, "Re: 'Totally Spam? It's Lubricated.'"

30. To make an extremely rough contemporary analogy, MUDs are like World of Warcraft, in which many players can work simultaneously but the game's world and play are largely in the hands of the administrators, and MOOs are like Second Life, in which players can construct spaces and objects within the game that other players can explore and use. The roughness of this analogy lies in the lack of programming liberty afforded to users of Second Life compared to those of MOOs, who could in many circumstances get under the hood of the world in interesting, experimental, and sometimes destructive ways.

31. See, for instance, Turkle, *Life on the Screen*; Molloy, "Public Literature: Narratives and Narrative Structures in Lambda MOO"; Dibbell, "A Rape in Cyberspace" and *My Tiny Life*.

32. Shaviro, *Doom Patrols*, 136.

33. Dibbell, "A Rape in Cyberspace." See also Dibbell, *My Tiny Life*, 19.

34. Hafner, *The Well*, 53.

35. Hess, *Yib's Guide to MOOing*, 29.

36. Dibbell, *My Tiny Life*, 100.

37. Under her alias "Sunny": Hess, *Yib's Guide to MOOing*, 321.

38. Dibbell, *My Tiny Life*, 280.

39. Dibbell, *My Tiny Life*, 19.

40. Dibbell, *My Tiny Life*, 97.

41. Chapman, *The Works of George Chapman*, 4.

42. Stephenson. "Mother Earth Mother Board," 95–161.

43. Stephenson, *Anathem*.

44. Stallman, "Why Schools Should Exclusively Use Free Software."

45. Lions, *Commentary on UNIX 6th Edition*.

46. Department of Defense,"DoD Internet Host Table."

47. Spatt,"Postel,Jon," 450. See also BBC News (no byline),"'God of the Internet' Is Dead."

48. Ryan, *A History of the Internet*, 33.

49. Crocker, "How the Internet Got Its Rules."

50. Crocker, RFC 3: "Documentation Conventions."

51. BBC News, "'God of the Internet' Is Dead."

52. Abbate, *Inventing the Internet*, 70.

53. Postel, RFC 706: "On the Junk Mail Problem."

54. Edwards, *The Closed World*, 111.

55. On Baran and early packet-switching theory, see Abbate, *Inventing the Internet*, 10–21.

56. Crocker, RFC 3: "Documentation Conventions."

57. Edwards, *A Vast Machine*, 25.

58. North and Iseli, *ARPANET News*, 5, 7, 16.

59. Kleinrock, interview by Petrie, "Len Kleinrock on the Origins of the Internet."

60. Walker, "MSGGROUP# 002 Message Group Status."

61. Crocker, "MSGGROUP# 004 Use of a Teleconferencing system, in place of Net Mail."

62. Chansler, "Re: Close, but No Cigar," as quoted in Brian Reid, "MSGGROUP# 506 Message headers: a note from the grass roots."

63. Postel, "MSGGROUP# 561 Comments on RFC 724."

64. Benkler, "Sharing Nicely," 273–358.

65. Pickering, *The Cybernetic Brain*, 17–22.

66. Bowker and Star, *Sorting Things Out*, 1999.

67. Van Vleck, "The History of Electronic Mail." The documentary filmmaker and essayist Errol Morris conducted several fascinating interviews with Van Vleck and others connected with the history of time sharing and email ("Did My Brother Invent E-Mail With Tom Van Vleck?," http://opinionator.blogs.nytimes.com/2011/06/19/did-my-brother-invent-e-mail-with-tom-van-vleck-part-one/).

68. Stallman, "MSGGROUP# 697 Some Thoughts about Advertising."

69. Hafner and Lyon, *Where Wizards Stay Up Late*, 207. Oddly, Crowther had been the subject of an earlier search on ARPANET, leading to a discussion of whether to have some sort of user-portable address—see Martin, "MSGGROUP# 546 ABSENTEE ADDRESSEES." A superb analysis of the genesis and meaning of Adventure can be found in Jerz, "Somewhere Nearby Is Colossal Cave."

70. "Polylogue" was coined by Austin Henderson, in "MSGGROUP# 522 Re: CONTENTS OF SUBJECT FIELDS"; "a select group of people" as described by Mark Crispin, "MSGGROUP# 696 in reply to Jake's message about advertising."

71. The description of the Quasar robot is quoted by Philip Carlton in a message forwarded to the MSGGROUP list: Nelson, "MSGGROUP# 569 Does it know about mail, too?"

72. Reid, "MSGGROUP# 614 Fake Robot: A Call for Help."

73. Stefferud and Farber, "MSGGROUP# 675 The Quasar Discussion."

74. As forwarded to the MSGGROUP list: Goodfellow, "MSGGROUP# 699 [THUERK at DEC-MARLBORO: ADRIAN@SRI-KL]." See also Templeton, "Reaction to the DEC Spam of 1978."

75. Kendall, "Community and the Internet," 310.

76. Deutsch, "MSGGROUP# 684 Re: The Quasar Discussion."

77. Hauben, "The Evolution of Usenet News." Note that the name of the SF-Lovers list is misspelled in this document as "Duffy," which confuses him with the architect Roger Duffy; Duffey's work on digests and the SAVE-LARGE-LISTS project is a very interesting early case of moderating large volumes of email communication.

78. Stallman, "MSGGROUP# 697 Some Thoughts about Advertising."

79. Kropotkin, "Law and Authority: An Anarchist Essay," 202.

80. Stallman, "MSGGROUP# 698 DEC Message [VERY TASTY!]."

81. McCarthy, "MSGGROUP# 692 Reaction."

82. Zellich, "MSGGROUP# 693 INOVATIONS IN ENGINEERING PUBLICATION."

83. Crispin, "MSGGROUP# 696 in reply to Jake's message about advertising."

84. Price and Verhulst, *Self-Regulation and the Internet*, 14.

85. Nelson, *Dream Machines*.

86. Pfaffenberger, "'If I Want It, It's Okay,'" 367.

87. Hauben and Hauben, "On the Early Days of Usenet."

88. Den Beste, "Trivia on the Net."

89. Pfaffenberger, "'If I Want It, It's Okay,'" passim.

90. Hayes, "An Alternative Primer on Net Abuse, Free Speech, and Usenet."

91. Hayes, "An Alternative Primer on Net Abuse, Free Speech, and Usenet"; note also the brief argument that "antispam zealotry" is leading to ISPs delivering inferior service.

92. Woodbury, as cited in Pfaffenberger, "'If I Want It, It's Okay,'" 380.

93. Furr, "Re: ARMM: ARMM: >>>>Ad Infinitum."

94. Wiener, "Nebraska letter."

95. "JJ"/Rob Noha, "HELP ME!!AA."

96. Webber, "FCC? U.S.Mail.? (Re: JJ's Revenge—Part II)."

97. Customer Service at Portal Communications, "JJ's Posting."

98. Customer Service at Portal Communications, "A Note From Portal Regarding the 'JJ' Incident."

99. I am indebted to Mario Biagioli for first suggesting this parallel.

100. Davis, *The Return of Martin Guerre*, 21.

101. Palmer, "Discordant Music," 5–62.

102. Hardy, *The Mayor of Casterbridge*, 366, 369.

103. For an interesting parallel case in violating privacy and collectively producing public shame online, see Liu, "Human Flesh Search Engine."

104. Johnson, "Due Process and Cyberjurisdiction," 334.

105. Johnson, "Due Process and Cyberjurisdiction," 332.

106. Fraser, "Viral Vigilantes."

107. This is obviously only a very brief summary of a fascinating subject. For a much fuller historical understanding, particularly of the central role of violence, see Rosenbaum and Sederberg, eds., *Vigilante Politics*; French, *The Virtues of Vengeance*; and Abrahams, *Vigilant Citizens*.

108. As cited in Forbes, *The Satiric Decade*, 207. For English instances of the charivari, see Thompson, "Rough Music: Le Charivari Anglais," 285–312, and "Rough Music Reconsidered," 3–26. For North American instances, see Palmer, "Discordant Music," 5–62.

109. Forbes, *The Satiric Decade*, 177.

110. Haugh II, "Re: HELP ME!!!"

111. Kleinpaste, "Re: C&S Have Declared War on the Net. How to Defend the Net?"

112. Abbate, *Inventing the Internet*, 185.

113. Pfaffenberger, "'If I Want It, It's Okay,'" 384.

114. Cliff Figallo, as quoted in Hafner, *The Well*, 55.

115. Turner, *From Counterculture to Cyberculture*, 73.

116. Rheingold, "What the WELL's Rise and Fall Tell Us about Online Community."

117. Hafner, *The Well*, 114–118.

118. Grossman, *Net.Wars*, 33.

119. Canter, interview by Feist, "The Father of Modern Spam Speaks."

120. Canter, as quoted in Moran, "The Day the Net Changed Forever."

121. Stivale, "Spam: Heteroglossia and Harassment in Cyberspace," 133–144.

122. Scurr, "Re: Green Card Lottery- Final One?"

123. Larson, "Re: Green Card Lottery- Final One?"

124. Nicholson, "Re: Green Card Lottery- Final One?"

125. Gillett, "bozo lawyers."

126. Cantillo, "Re: Green Card Lottery- Final One?"

127. Canter, interview by Feist, "The Father of Modern Spam Speaks."

128. Friedman, "Re: Green Card Lottery- Final One?"

129. Kilna, "Re: Green Card Lottery- Final One?"

130. Lewis, "Sneering at a Virtual Lynch Mob."

131. Ackerman, "Re: Green Card Lottery- Final One?"

132. Flynn, "'Spamming' on the Internet."

133. Flynn, "'Spamming' on the Internet."

134. Canter and Siegel, *How to Make a Fortune on the Information Superhighway*, 187.

135. Werry, "Imagined Electronic Community."

136. Canter and Siegel, *How to Make a Fortune on the Information Superhighway*, 12, 188. Thatcher's famous quote can be found in Thatcher, interview by Douglas Keay, "Aids, Education and the Year 2000!"

137. Joselit, *Feedback: Television against Democracy*, 105.

138. Grossman, *Net.Wars*, 19.

139. Rheingold, "The Tragedy of the Electronic Commons."

140. Canter, interview by Feist, "The Father of Modern Spam Speaks."

141. Grossman, *Net.Wars*, 25.

142. Lloyd, "Re: Green Card Lottery- Final One?"

2 MAKE MONEY FAST

1. Hofstadter, *Gödel, Escher, Bach*, 286.

2. Latour, "Can We Get Our Materialism Back, Please?," 138–142.

3. McWilliams, *Spam Kings*.

4. "This term is derived from a skit performed on the British television show *Monty Python's Flying Circus*, in which the word 'spam' is repeated to the point of absurdity in a restaurant menu." *CompuServe v. Cyber Promotions, Inc.*

5. As discussed on NANAE: Leader, "Re: I'M OUT!," in news.admin.net-abuse.email. See also Kanaley, "Sanford Wallace, the Spam King, Abdicates and Apologizes Online."

6. Fuller and Goffey, "Toward an Evil Media Studies," 152.

7. Fuller and Goffey, "Toward an Evil Media Studies," 152.

8. Boyle, "Spam Hits the House of Representatives."

9. Lewis, "Protest Halts E-Mail 'Spam.'"

10. Lane, *Obscene Profits*, 154–155.

11. Lane, *Obscene Profits*, 158.

12. Wolcott, "You Call It Spam, They Call it a Living."

13. Brunker, "In the Trenches of the 'Spam Wars.'"

14. Wolcott, "You Call It Spam, They Call it a Living."

15. McWilliams, *Spam Kings*, passim. See also Moser, "Return of the 'Kosher Nazi.'"

16. Fitzgerald, "AOL Gives Up Treasure Hunt."

17. Scoblionkov, "Senate Embraces Spam Bill."

18. Angwin, *Stealing MySpace*, 23.

19. On Jason Heckel: Oman, "Washington Supreme Court Upholds State Antispamming Law," 931–937. On Davis Hawke: McWilliams, *Spam Kings*, 89. On Email America: The Virtual Magistrate (news release), "Virtual Magistrate Issues Its First Decision," 343–345.

20. "Who Is Premier Services," premeir-marketing.htm, beyond-enemy-lines archive, n.d.

21. Lialina, "A Vernacular Web 2," 58–69.

22. Deekoo, "editorial note" in "Lets Get Brutal!," http://deekoo.net/peeves/spam/spammers/premiere/brutal.htm.

23. Hilderbrand, *Inherent Vice*, 66–71.

24. "Rodona-and-DA.txt," beyond-enemy-lines archive, September 7, 1999.

25. "Aspen-and-Shannon.txt," beyond-enemy-lines archive, March 31, 2000.

26. "Rodona-and-Andy.txt," beyond-enemy-lines archive, September 7, 1999.

27. "Aspen-and-Shannon.txt," beyond-enemy-lines archive, May 22, 2000.

28. "Aspen-and-Shannon.txt," beyond-enemy-lines archive, March 31, 2000.

29. "Rodona-Server-and-Ken.txt," beyond-enemy-lines archive, September 14, 1999, and September 20, 1999.

30. "Rodona-Server-and-Ken.txt," beyond-enemy-lines archive, September 20, 1999.

31. McWilliams, *Spam Kings*, 222.

32. "mk590," "AOL for free?" in alt.2600, January 28, 1996.

33. "Aspen-and-Shannon.txt," beyond-enemy-lines archive, May 20, 2000.

34. "Aspen-and-Shannon.txt," beyond-enemy-lines archive, May 22, 2000.

35. "Aspen-and-Shannon.txt," beyond-enemy-lines archive, May 22, 2000.

36. "Rodona-and-Al.txt," beyond-enemy-lines archive, June 8, 1999.

37. "Rodona-Razzle-and-GateKeeper.txt," beyond-enemy-lines archive, January 18, 2000.

38. "Rodona-Razzle-and-GateKeeper.txt," beyond-enemy-lines archive, February 9, 2000.

39. "Rodona-Server-and-Ken.txt," beyond-enemy-lines archive, September 14, 1999.

40. "Rodona-Razzle-and-Dave-Gosse.txt," beyond-enemy-lines archive, November 15, 1999.

41. The "zealous little anti": "Aspen-and-Shannon.txt," beyond-enemy-lines archive, April 8, 2000. The Wizard: Lewis, "Protest Halts E-Mail 'Spam.'" "Wild-eyed zealots": Flynn, "'Spamming' on the Internet."

42. "Rodona-Razzle-and-Charlie.txt," beyond-enemy-lines archive, passim.

43. "M-ANTI04.txt," beyond-enemy-lines archive.

44. "Rodona-Razzle-and-Charlie.txt," beyond-enemy-lines archive, April 12, 2000.

45. "Aspen-and-Shannon.txt," beyond-enemy-lines archive, April 6, 2000.

46. "Rodona-Razzle-and-Charlie.txt," beyond-enemy-lines archive, May 6, 2000.

47. "Rodona-Razzle-and-Charlie.txt," beyond-enemy-lines archive, April 9, 2000.

48. "Rodona-and-Boomer.txt," beyond-enemy-lines archive, June 9, 1999.

49. "Rodona-Server-and-Ken.txt," beyond-enemy-lines archive, September 14, 2000.

50. Rodona Garst, "Stock," to Mark Rice, October 2, 1999, in "Pump-N-Dump.htm," beyond-enemy-lines archive.

51. "Rodona-Server-and-Ken.txt," beyond-enemy-lines archive, September 20, 1999.

52. FTC (news release), "FTC Launches Crackdown on Deceptive Junk E-mail," February 12, 2002.

53. SEC Administrative Proceeding File No. 3-10843, July 24, 2002.

54. Tim Skirvin, "What Is the Format of a Cancel Message," in "Cancel Messages" FAQ, sec. II.C, ver. 1.75.

55. Kostakis, "Identifying and Understanding the Problems of Wikipedia's Peer Governance."

56. Webb, "COMMUN—THE FEMINIZATION OF CYBERSPACE."

57. Maltz, "Customary Law and Power in Internet Communities." On redaction of Soviet political media, see King, *The Commissar Vanishes.*

58. Hayes, "The USENET Site of Virtue FAQ."

59. Coleman, "Old and New Net Wars over Free Speech."

60. Grossman, *Net. Wars*, 74–80.

61. Some of the complex distinctions inherent in spam, "trash," "junk," and "waste" are considered in Galloway and Thacker, "On Narcolepsy," 145–147; Gansing, "The Production of Waste."

62. Latour, *Aramis, or the Love of Technology*, 59.

63. "sine nomine," "proposed press release, 2nd draft," in news.admin.misc, June 6, 1994.

64. Latour, *Reassembling the Social*, 185.

65. Johnston, *Technological Turf Wars*, 1.

66. Johnston, *Technological Turf Wars*, 18.

67. An in-depth analysis of the complex history of the "virus" as metaphor and as program can be found in Parikka, *Digital Contagions.*

68. "Rodona-Razzle-and-Dave-Gosse.txt," beyond-enemy-lines archive, November 15, 1999.

69. Singel, "Curtain Call for Junk-Fax Blaster."

70. Vixie and Nicholas, "How to Sue MAPS."

71. U.S. House of Representatives, *Netizens Protection Act of 1997.*

72. Templeton, "Problems with H.R. 1748 (the 'Smith Bill')."

73. Wu, *The Master Switch*, 104–107.

74. Casey, *ISP Liability Survival Guide*, 139.

75. Brand, *The Media Lab*, 62.

76. Burk, "The Trouble with Trespass," 27–56.

77. I am in the process of working on a long article about the history of law enforcement, governance, and antispam—including the famous "Fridge" in the FTC, which received 300,000 spam messages a day and provided a database of evidence for cases brought by groups across many agencies and departments—that can supplement this necessarily brief look at the federal side of the antispam effort.

78. The discussion group alt.spam, started in 1990 as a place to discuss the lunch meat and engage in endless repetitions of the Monty Python sketch, shifted during mid-1990s to a collection of email-related outrages, but the conversation there was far less coherent and solution-focused than that of NANAE. It was an etymological accident, not a deliberately formed activist gathering.

79. Mueller, *Ruling the Root*, 162.

80. Edwards, "Re: ABUSE: CHAG," in news.admin.net-abuse.email, news.admin.net-abuse.misc, July 15, 1996.

81. Hollis and others, "alt.spam FAQ or 'Figuring out fake E-Mail & Posts,'" rev. January 30, 2005.

82. 15 U.S.C. 7701 et seq., Public Law No. 108–187, S.877.

83. Lessig, *Code and Other Laws of Cyberspace*, 266.

84. McWilliams, *Spam Kings*, 270.

85. Graham, "Spam is Different."

86. McWilliams, *Spam Kings*, 139.

87. McWilliams, *Spam Kings*, 274.

88. Hansell, "Spam Sender Settles Case in New York."

89. Graham, "So Far, So Good."

90. Dixon, "'I Will Eat Your Dollars.'"

91. Scam-baiting also has a theme of barely disguised racism, as discussed by Graham Parker in one of the pamphlets ("419 (occasional 420): Reston, Virginia & Lagos, Nigera, 2005") in his collection *Fair Use (Notes from Spam)*—the manifest, gleeful delight, and sententious self-righteousness expressed by comparatively rich and powerful Westerners at their computers as they toy with and mock impoverished West Africans, including sending them into very dangerous situations.

92. Train, "The Spanish Prisoner."

93. "An Old Swindle Revived," *New York Times*, March 20, 1898.

94. Think of the famous "dollar auction" experiments developed for economics theory by Martin Shubik, in which players, bidding for a dollar and constrained to pay whatever they bid, often end up bidding far above the value of that dollar

in their effort to recoup the cost of their bids. See Shubik, "The Dollar Auction Game."

95. Smith, *A Culture of Corruption.*

96. Smith, *A Culture of Corruption*, 38.

97. Hafez Ringim, as quoted in Kaplan, "A Land Where Con Is King."

98. Burrell, "Problematic Empowerment," 15–30.

99. Armstrong, "'Sakawa' Rumours."

100. McDermott, "US Postal Service Is Cracking Down on Nigerian Scams."

101. *Adler v. Republic of Nigeria*, 219 F.3d 869 (9th Cir. 2000), case no. 98-55456.

102. Glenny, *McMafia*, 166–167.

103. Glenny, *McMafia*, 161–164.

104. For Royal Dutch Shell's manipulation of the Nigerian government, see Smith, "WikiLeaks Cables."

105. Jones, "The Robot-Readable World."

106. Gibson, *Zero History*, 345.

107. This echoes the groundbreaking and underappreciated work of Jakob von Uexküll, who founded the field of biosemiotics with ideas such as the *Umwelt* (roughly, "surrounding world") of a creature—that different kinds of entities can inhabit radically different forms of the same environment, with the same object having a different "functional coloring" (a more nuanced form of affordance theory) for each, depending on their needs, bodies, and senses. See, for instance, von Uexküll, *A Foray into the Worlds of Animals and Humans.*

108. Jones, "Sensor-Vernacular"; Bridle, "The New Aesthetic."

109. McNeil, "The New Aesthetic," March 14, 2012.

110. Newman, "Poor Penmanship Spells Job Security for Post Office's Scribble Specialists."

111. comScore, "comScore Releases January 2011 U.S. Search Engine Rankings" (news release), February 11, 2011.

112. On coverage bias: Call et al., "Bias in Telephone Surveys that Do Not Sample Cell Phones," 355–364.

113. Atwood, "The Elephant in the Room."

114. Optify Research, "The Changing Face of SERPs"; Chitika Insights, "The Value of Google Result Positioning."

115. See also the fascinating meditation on the history—on the historical nature—of web pages and their markup in Gitelman, *Always Already New*, chapter 4.

116. Sullivan, "Death of a Meta Tag."

117. Pringle et al., "What Is a Tall Poppy among Web Pages?," 369–377.

118. Human cultural analogies for this split text are interesting: allegorical images, narratives with different meanings for those with different interpretative backgrounds, anamorphosis in painting with different images revealed to specialized tools. But these examples don't speak to the automated process of making biface texts, which is interesting to consider at greater length as we develop a more algorithmic culture.

119. A very thought-provoking study of the use of robots.txt files to censor and conceal: Elmer, "Robots.txt: The Politics of Search Engine Exclusion," 217–227.

120. Brin and Page, "Anatomy of a Large-Scale Hypertextual Web Search Engine," 107–117.

121. John Battelle has chronicled the many points at which the Google we know today very nearly became something entirely different from its present form, whether by licensing its algorithm or simply never coming together as a coherent corporation. See Battelle, *The Search*.

122. An interesting question can be raised in relation to this innovative development: why is it innovative—or rather, why wasn't such a system already normal? Citation analysis has been a common tool in social science since the Science Citation Index® began in 1963 (with the original outline of such an idea in 1955, in Eugene Garfield's "Citation Indexes for Science"). Some of the conceptual elements of such a system can be found as far back as the 1873 publication of the first *Shepard's Citations*, which mapped legal citations. So why did such an approach only appear with "third-generation" search, rather than at the outset? Was it the conceptual understanding of the web in the initial years of browser development—that is, did the web not look "social," at first, in the way a citation system does? Was it the separation of different disciplines within academia and business, such that the engineers, programmers, and entrepreneurs of the early web did not have a range of theories with which to analyze their project? See Garfield, "Citation Indexes for Science," 108–111.

123. Kumar et al., "Trawling the Web," 1481–1493.

124. Brin and Page, "Anatomy of a Large-Scale Hypertext Search Engine," 2.

125. Brin and Page, "Anatomy of a Large-Scale Hypertext Search Engine," 4.

126. Brin and Page, "Anatomy of a Large-Scale Hypertext Search Engine," 4.

127. Brin and Page, "Anatomy of a Large-Scale Hypertext Search Engine," 4.

3 THE VICTIM CLOUD

1. In the case of spam distributed by the networks of compromised computers called "botnets," the spam production rates decline at night when people turn their computers off.

2. Metsis et al., "Spam Filtering with Naive Bayes."

3. "In place of seven thousand thirteen, he would say (for example) Máximo Perez; in place of seven thousand fourteen, The Train; other numbers were Luis Melián Lafinur, Olimar, Brimstone, Clubs, The Whale, Gas, The Cauldron, Napoleon, Agustín de Vedia. In lieu of five hundred, he would say nine," as Borges put it in "Funes the Memorious." It recalls Gödel numbering, as well, which encodes symbols and formulas expressed in formal language with unique natural numbers, making them manipulable and computable in interesting ways. See Borges, *Ficciones*.

4. There are also curious technical difficulties with the use of the number-substitution approach: "The majority of filters we have evaluated exhibit pathologies on the PU obfuscated corpora," write Cormack and Lynam of number-substitution—the profile of messages that have undergone the obfuscation process is different enough that filters built for natural language materials will not work correctly, with corresponding difficulties the other way around. See Cormack and Lynam, "Spam Corpus Creation for TREC."

5. For a look at Ritter's use of his own body as the experimental apparatus, see Zielinski, *Deep Time of the Media*, 177.

6. Rennie, "ifile."

7. Klimt and Yang, "The Enron Corpus."

8. Federal Energy Regulatory Commission, "Western Energy Markets: Major Issuance on March 26, 2003—Information Released in Investigation," http://web.archive.org/web/20031011133628/http://ferc.gov/industries/electric/indus-act/wem/03-26-03-release.asp.

9. William W. Cohen, a computer scientist based at Carnegie Mellon, promoted the Enron materials as a corpus and made them available on the web, and he in turn credits Leslie Kaelblin, at MIT and Melinda Gervasio at SRI, for purchasing the entire dataset from the FERC and working on integrity problems that made it initially problematic as a corpus. In the DIY spirit of Jason Rennie and the hackers we will be looking at shortly, the reader is welcome to explore the corpus themselves: all 400 megabytes of data are available for download from http://www.cs.cmu.edu/~enron/enron_mail_030204.tar.gz.

10. Klimt and Yang, "The Enron Corpus," 2004.

11. Recent dominant framework: Cormack and Lynam, "Spam Corpus Creation for TREC." "Repeatable": Cormack and Lynam, "TREC 2005 Spam Track Overview."

12. Cormack and Lynam, "Spam Corpus Creation for TREC," 1.

13. Cormack and Lynam, "Spam Corpus Creation for TREC," 2.

14. Cormack and Lynam, "TREC 2005 Spam Track Overview."

15. Sahami et al., "A Bayesian Approach to Filtering Junk Email."

16. Graham, "A Plan for Spam."

17. Wiener, *Cybernetics*, 27.

18. Graham, "A Plan for Spam."

19. The "naïve" in naïve Bayes means that different factors, such as different words or tokens, are considered separately in assessing probability. They do not affect each other, and the probability associated with one word does not alter the probability associated with another.

20. For a theoretical study of the problem of why naïve Bayes works as well as it does: "Its competitive performance in classification is surprising, because the conditional independence assumption on which it is based, is rarely true in real-world applications. An open question is: what is the true reason for the surprisingly good performance of naive Bayes in classification?," see Zhang, "The Optimality of Naive Bayes," 1.

21. Rennie, "ifile." For ifile's original announcement, see Rennie, "ifile README 0.1a." Much of the theory underlying ifile can be found in his thesis: Rennie, "Improving Multi-class Text Classification with Naive Bayes.". ifile, written in C and Perl, has the theoretical skeleton of many programs later in this history and provides a useful illustration of the specific intricacies of building a Bayesian filter.

22. This estimate, compiled by Ferris Research, is quoted repeatedly throughout the antispam literature of the 2000s. See, for instance, Cukier et al., "Genres of Spam," 2. That deletion interval is multiplied out to various millions of user-hours, depending on the metrics of analyst, and is often considered as cost in salaried working time.

23. Graham, "Better Bayesian Filtering."

24. Sahami et al., "A Bayesian Approach to Filtering Junk Email."

25. Cardwell, "SpamAssassin 2010 Bug."

26. Graham, "Better Bayesian Filtering."

27. The full explanation of this critique would take too much time in this context; the interested reader is directed to Peters's post: "Re: [Python-checkins] python/nondist/sandbox/spambayes GBayes.py,1.7,1.8," http://mail.python.org/pipermail/python-dev/2002-August/028216.html.

28. Graham, "Plan for Spam FAQ"; Graham, "Will Filters Kill Spam?"

29. Graham, "A Plan for Spam."

30. Graham, "Will Filters Kill Spam?"

31. Graham, "So Far, So Good."

32. This is known as Zipf's Law, after the philologist and linguist George K. Zipf (the kind of thinker would could be a "University Lecturer" at Harvard, meaning that he could teach any subject). See Zipf, *Human Behavior and the Principle of Least Effort*; *The Psycho-Biology of Language*.

33. Stephenson, "Cryptonomicon Spam."

34. Kafka, *The Metamorphosis and Other Stories*, 177.

35. For a consideration of the reverse Turing test, see Hofstadter, "The Turing Test: A Coffeehouse Conversation."

36. Alder, "America's Two Gadgets, 124–137; *The Lie Detectors*; Kinsmen, "'Character Weakness' and 'Fruit Machines,'" 133–161.

37. Turing, "Computing Machinery and Intelligence," 433–460.

38. For a deeper study of this noticeably gendered setup, see Genova, "Turing's Sexual Guessing Game."

39. Stross, "It's Made Out of Meat."

40. Preston, *The Wild Trees*, 215.

41. Carter, "From Drawings to Sets," 1:00.

42. White, *Identity and Control*, 4.

43. Graham, "A Plan for Spam."

44. For 2006 numbers, see Phifer, "2006 MSSP Survey, Part 6." For 2011 numbers, see Messaging Anti-Abuse Working Group, "Email Metrics Report."

45. McWilliams, *Spam Kings*, 173.

46. Graham, "Will Filters Kill Spam?"

47. Graham, "Plan for Spam FAQ."

48. "Terra," http://tyler-city.blogspot.com. This splog has subsequently been taken down by Blogger.

49. Fetterly et al., "Spam, Damn Spam, and Statistics."

50. The "via" link has interesting parallels with other social practices: the cab driver who recommends a relative's hotel and then adds, "Tell them I sent you," or the advertising circular that specifies that the shopper mention where he or she saw the advertisement for a special discount.

51. Beckett, as quoted by Harris, *William Burroughs and the Secret of Fascination*, 158.

52. Carr, "The 'Quick Excerpt' Splog."

53. Hartcup, *Camouflage*, 71.

54. In this regard, we should also mention wiki spam: when spammers exploit publically editable wikis to create new pages full of links that, again, look to a search engine spider like an endorsement of the linked site. I once ran an academic wiki on a university server that was pirated in this fashion; the links, largely in Chinese, included a granite wholesaler in Qingdao; a bunch of Chinese ISO-9000/9001 consulting firms; a media-piracy, P2P, and instant message front-end called netbai; and a couple of gold-farming outfits for buying video game resources.

55. Google Inc., Securities Exchange Commission Form 10-Q, 23.

56. DataPresser, "Pricing & Signup."

57. "Google only pays you adsense revenue because they are making money," writes "The Junk Man," a manipulator of AdSense revenue, "and all they care about is keeping their customers happy. If your content is pin point and you are capturing the proper audience you are doing google's advertisers a favor. . . . That is what googles [sic] business revolves around!" The Junk Man, "MNS: Bury the Hatchet."

58. Anderson, "From Indymedia to Demand Media."

59. Roth, "The Answer Factory."

60. Anderson, "Deliberative, Agonistic, and Algorithmic Audiences."

61. Anderson, "Deliberative, Agonistic, and Algorithmic Audiences," 542.

62. Carlson, "LEAKED: AOL's Master Plan."

63. Fishkin, "Linkbaiting for Fun and Profit."

64. Sibona and Walczak, "Unfriending on Facebook."

65. Marwick, "Status Update."

66. Amazon hosts automated email spam services as well, through its Elastic Computing Cloud (EC2) platform, which rents room and bandwidth on Amazon's servers to small business—Amazon has much more server capacity than it can use

at any given time—and turns out to be a fantastic platform for running spam operations on the cheap, payable with stolen credit cards. See Krebs, "Amazon."

67. Perez, "Amazon's Mechanical Turk Used for Fraudulent Activities."

68. Jordan, "Social Bookmarking My Web Site," Amazon Mechanical Turk task. (See also a screen capture of the task in Evans, "Ethically Improper Use of Mechanical Turk?")

69. Discussion was taking place at the following URL: http://www.blackhatworld.com/blackhat-seo/craigslist-other-classified-ads-sites/13487-craigslist-phone-verification-system.html#post121327. It has subsequently been taken down by the site in question, following bad publicity.

70. http://www.blackhatworld.com/blackhat-seo/craigslist-other-classified-ads-sites/13487-craigslist-phone-verification-system-4.html#post163014.

71. Whoriskey, "Digital Deception."

72. Prasad, "Microsoft's CAPTCHA Revolutions Busted by Spammers."

73. Krebs, "Virtual Sweatshops Defeat Bot-or-Not Tests." For a thorough (and fascinating) study of the economics of solving CAPTCHAs in bulk, see Motoyama et al., "Re:CAPTCHAs—Understanding CAPTCHA-Solving Services in an Economic Context."

74. Kelly, "Spammer AI."

75. Kremen, "'Emergent' Images to Outwit Spambots."

76. Motoyama et al., "Re:CAPTCHAs—Understanding CAPTCHA-Solving Services in an Economic Context," 442.

77. Brunner, *The Shockwave Rider*, 252.

78. Shoch and Hupp, "The 'Worm' Programs."

79. Shoch and Hupp, "The 'Worm' Programs," 172.

80. Shoch and Hupp, "The 'Worm' Programs," 173.

81. SecureList, "Email-Worm.Win32.Mydoom.a."

82. Leonard, *Bot*, 106.

83. Leonard, *Bot*, 140–144.

84. Pappalardo and Messmer, "Extortion via DDoS on the Rise." The specific nature of the Mydoom worm's project shows just how socially sophisticated these technical worms could be: the DDoS attack was itself a blind—a feint. SCO Inc. had been involved in a series of incendiary high-profile lawsuits against the Linux community related to arcane copyright issues; the idea that an outraged Linux

hacker had written a criminal worm specifically to take them out was a spectacular story in the technology press. (The worm included a similar attack against Microsoft on a different date, further bolstering the open-source-gone-wrong argument.) When the date arrived, only about a fifth of the infected machines delivered on the attack—for various reasons mostly related to buggy software and implementation (see Sophos Threat Analyses, "W32/MyDoom-B")—but in all the excitement around the planned DDoS, the worm's backdoor function, and therefore its capacity to use the infected computer to generate spam and export sensitive data to the controller, went largely unnoticed. See Grow and Bush, "Hacker Hunters."

85. Pansters, "mydoom.exe decyphering?"

86. Ianelli et al., "Botnets as a Vehicle for Online Crime."

87. Finjan Software, "Finjan Discovers Compromised Business and Customer Data of 40 Top-Tier Global Businesses."

88. Franklin et al., "An Inquiry into the Nature and Causes of the Wealth of Internet Miscreants."

89. Franklin et al., "An Inquiry into the Nature and Causes of the Wealth of Internet Miscreants."

90. Ianelli and Hackworth, "Botnets as a Vehicle for Online Crime."

91. This and other exchanges were observed on the public IRC channel #CC. power, February 23, 2009.

92. Holt and Lampke, "Exploring Stolen Data Markets On-Line"; Holt, "Examining Costs, Benefits, and Economics in Malware and Carding Markets."

93. Kreibich et al., "On the Spam Campaign Trail."

94. They have an intricate system of dictionaries from which to draw their variant elements, such as linksh (IP addresses of proxy bots designed to trick the message recipient into executing a file), names for email From: fields ("steven88"), pharma_links (domains for pharmaceutical ads, such as "http://iygom.tryyoung.cn/?625112501432"), and ronsubj (subject lines, as in "Best job for you").

95. Kreibich et al., "On the Spam Campaign Trail."

96. Borges, *Ficciones*, 83.

97. About Cutwail, see Rowe, Wood, and Reeves, "How the Public Views Strategies Designed to Reduce the Threat of Botnets," 339.

98. Enright et al., "Storm: When Researchers Collide."

99. Chesterton, *The Man Who Was Thursday*.

100. Gibson, *Neuromancer*, 259.

101. Enright et al., "Storm: When Researchers Collide."

102. Tung, "Storm Worm: More Powerful than Blue Gene?"

103. Greene, "Storm Worm Strikes Back at Security Pros."

104. Dagon, Zou, and Lee, "Modeling Botnet Propagation Using Time Zones."

105. Gady, "Africa's Cyber WMD."

106. Williamson, "Carpet Bombing in Cyberspace."

107. Nazario, "Estonian DDoS Attacks."

108. A comparable situation could be found in the Marshall Islands in the western Pacific, whose sole Internet provider was taken down on June 24, 2008, by a flood of automated emails; overwhelming this lone point of communication was able to effectively cut off the 55,000 inhabitants of the islands. Leyden, "Spam DDoS Assault Cuts Off South Pacific State."

109. Schwartz, "Preparing for a Digital Pearl Harbor."

110. Davis, "Hackers Take Down the Most Wired Country in Europe."

111. Anderson, "Massive DDoS Attacks Target Estonia; Russia Accused."

112. Williamson, "Carpet Bombing in Cyberspace."

113. Wall, *Cybercrime*, 144.

114. Greenberg, "Reformed Hacker Looks Back"; Littman, *The Fugitive Game*, 229.

115. Johnston, *Technological Turf Wars*, 58.

116. Johnston, *Technological Turf Wars*, 57.

117. McColo Hosting Solutions, "About Company."

118. Liu, interview by Lovink, "net critique."

119. See, for example, Gilman, Weber, and Goldhammer, *Deviant Globalization*; Neuwirth, *Stealth of Nations*; and Glenny, *McMafia*.

120. Williams, "The Largest Cloud in the World."

121. Hansen, "Conversations with a Blackhat."

122. On the Rustock situation, see Symantec.cloud, "Global Spam Drops as Rustock Botnet Is Dismantled."

123. Anderson et al., "Spamscatter."

124. Levchenko et al., "Click Trajectories," 1.

125. Specifically, the Unlawful Internet Gambling Enforcement Act of 2006, 31 USC §§ 5361–5367.

126. On disguising transactions, see Claburn, "Internet Poker Sites Seized." On the response from other WTO member nations, see Reuters, "WTO Confirms U.S. Loss in Internet Gambling Case"; Pfanner and Timmons, "U.K. Seeks Global Rules for Online Gambling."

127. On Twitter spambots, see Yardi et al., "Detecting Spam in a Twitter Network." On "likejacking," see Kharif, "'Likejacking': Spammers Hit Social Media."

CONCLUSION

1. This description refers to AOHell, an astonishing nightmare of a system built largely on the relatively easy language Visual Basic that took advantage of AOL's technical flaws to run what amounted to a covert file-sharing haven and communications network using AOL's hardware. With the right commands, trivially acquired, you could contact automated programs that would provide you with pirated software, music, and porn (the major components of the adolescent's Internet), use AOL for free, create fake accounts, send people messages that would take them offline, steal passwords—and ruin chat conversations with annoying pranks and deluge others with automated mail, both of which are part of pre-1994 "spamming." You could turn AOL, whose whole business model was built on being a kind of gated community that was safer than the open Internet, into a reservoir of IP violations and sleaze as uncouth as any BBS run out of the closet of an anarchist squat. The documentation for the 3.0 version of AOHell gives a good sense of the whole amazing affair: Da Chronic, "AOHell v3.0 Rage Against The Machine."

2. In the analysis of some theorists, in fact, spam is a pure expression of the network itself: the architecture rendered explicit. Alessandro Ludovico considers spam as only one instance in a long history of projects, from traveling salesmen to personalized bulk postal mail to eye-catching billboards, designed to interfere with our thoughts and provoke us into some form of consumer desire—people on networked computers were no more exempt from this business then those who happen to be home from hearing the peddler's knock ("Spam, the Economy of Desire"). Other analyses draw on work following Ulrich Beck's concept of the risk society, such as Joost van Loon's, reinterpreted by Jussi Parikka and Tony Sampson as a step toward an analysis where "bad objects" are nothing more than the "manufactured side effects" of the good; they can never be wholly purged (unless we are willing to give up the goods which enable them) but can be managed, mitigated, their potential harm contained. (See Parikka and Sampson, "Part II: Bad Objects," 101.) For Geert Lovink, spam and other points of failure like identity theft are already present from the outset, "not accidental mistakes"

but "constitutional elements of yesterday's network architecture" ("The Principle of Notworking").

3. Benkler, *The Wealth of Networks*, 33. See also Litman, "Stealing and Sharing."

4. Kelly, "Networked Books and Networked Reading." See in particular the work of Philip Parker, who has "written" hundreds of thousands of books (Cohen, "He Wrote 200,000 Books"). See also his original patent filing: Philip M. Parker, "Method and Apparatus for Automated Authoring and Marketing," U.S. Patent 7,266,767, issued September 4, 2007.

BIBLIOGRAPHY

Abbate, Janet. *Inventing the Internet.* Cambridge, Mass.: MIT Press, 1999.

Abrahams, Ray. *Vigilant Citizens: Vigilantism and the State.* Cambridge: Polity Press, 1999.

Ackerman, Brad. "Re: Green Card Lottery- Final One?" In alt.internet.services, alt.tasteless, alt.cyberpunk, alt.culture.internet, news.admin.policy, April 23, 1994.

Adler v. Republic of Nigeria, 219 F.3d 869 (9th Cir. 2000), case no. 98–55456.

Agar, Jon. *The Government Machine: A Revolutionary History of the Computer.* Cambridge, Mass.: MIT Press, 2003.

Akera, Atsushi. *Calculating a Natural World: Scientists, Engineers, and Computers During the Rise of U.S. Cold War Research.* Cambridge, Mass.: MIT Press, 2007.

Alder, Ken. "America's Two Gadgets: Of Bombs and Polygraphs." *Isis* 98 (1) (March 2007): 124–137.

Alder, Ken. *The Lie Detectors: The History of an American Obsession.* Lincoln: University of Nebraska Press, 2009.

Anderson, C. W. "Deliberative, Agonistic, and Algorithmic Audiences: Journalism's Vision of its Public in an Age of Audience Transparency." *International Journal of Communication* 5 (2011): 529–547.

Anderson, C. W. "From Indymedia to Demand Media: Journalism's Visions of Its Audience and the Horizons of Democracy." In *The Social Media Reader*, ed. Michael Mandeberg, 77–96. New York: New York University Press, 2012.

Anderson, D. S., et al. "Spamscatter: Characterizing Internet Scam Hosting Infrastructure." *Proceedings of 16th USENIX Security Symposium* (2007).

Anderson, Nate. "Massive DDoS Attacks Target Estonia; Russia Accused." Ars Technica, May 2007. http://arstechnica.com/security/2007/05/massive-ddos-attacks-target-estonia-russia-accused/.

Angwin, Julia. *Stealing MySpace: The Battle to Control the Most Popular Website in America*. New York: Random House, 2009.

Armstrong, Alice. "'Sakawa' Rumours: Occult Internet Fraud and Ghanaian Identity." UCL working paper no. 8/2011.

Atwood, Jeff. "The Elephant in the Room: Google Monoculture." Coding Horror (blog), February 9, 2009. http://www.codinghorror.com/blog/2009/02/the-elephant-in-the-room-google-monoculture.html.

Battelle, John. *The Search: How Google and Its Rivals Rewrote the Rules of Business and Transformed Our Culture*. New York: Portfolio, 2005.

BBC News. "'God of the Internet' Is Dead." BBC News, October 19, 1998. http://news.bbc.co.uk/2/hi/science/nature/196487.stm

Benkler, Yochai. "Sharing Nicely: On Shareable Goods and the Emergence of Sharing as a Modality of Economic Production." *Yale Law Journal* 114 (2) (November 2004): 273–358.

Benkler, Yochai. *The Wealth of Networks: How Social Production Transforms Markets and Freedom*. New Haven: Yale University Press, 2006.

Borges, Jorge Luis. *Ficciones*. New York: Grove Press, 1962.

Bowker, Geoffrey C., and Susan Leigh Star. *Sorting Things Out: Classification and Its Consequences*. Cambridge, Mass.: MIT Press, 1999.

Boyle, Alan. "Spam Hits the House of Representatives." ZDNet, October 5, 1999. http://www.zdnet.com/news/spam-hits-the-house-of-representatives/103422.

Brand, Stewart. *The Media Lab: Inventing the Future at MIT*. New York: Penguin, 1988.

Bridle, James. "The New Aesthetic." RIG London (blog), May 6, 2011. http://www.riglondon.com/blog/2011/05/06/the-new-aesthetic/.

Brin, Sergey, and Lawrence Page. "The Anatomy of a Large-Scale Hypertextual Web Search Engine." *Computer Networks & ISDN Systems* 30 (1–7) (April 1, 1998): 107–117.

Brunker, Mike. "In the Trenches of the 'Spam Wars.'" MSNBC.com, August 7, 2003. http://www.msnbc.msn.com/id/3078650/ns/technology_and_science-security/t/trenches-spam-wars/.

Brunner, John. *The Shockwave Rider*. New York: Ballantine, 1976.

Burk, Dan. "The Trouble with Trespass." *Journal of Small and Emerging Business Law* 4 (2000): 27–56.

Burrell, Jenna. "Problematic Empowerment: West African Internet Scams as Strategic Misrepresentation." *Information Technology and International Development* 4 (4) (Fall/Winter 2008): 15–30.

Bygrave, Lee A., and Jon Bing. *Internet Governance: Infrastructure and Institutions*. Oxford: Oxford University Press, 2009.

Call, K. T., et al. "Bias in Telephone Surveys that Do Not Sample Cell Phones: Uses and Limits of Poststratification Adjustments." *Medical Care* 49 (4) (April 2011): 355–364.

Canter, Laurence. Interview by Sharael Feist, "The Father of Modern Spam Speaks." *CNET News*, March 26, 2002. http://news.cnet.com/2008-1082-868483.html.

Canter, Laurence, and Martha Siegel. *How to Make a Fortune on the Information Superhighway: Everyone's Guerrilla Guide to Marketing on the Internet and Other On-Line Services*. New York: HarperCollins, 1994.

Cantillo, Brandon A. "Re: Green Card Lottery- Final One?" In alt.internet.services, alt.tasteless, alt.cyberpunk, alt.culture.internet, news.admin.policy, April 13, 1994.

Cardwell, Mike. "SpamAssassin 2010 Bug." Grepular (blog), January 1, 2010. https://grepular.com/SpamAssassin_2010_Bug.

Carlson, Nicholas. "LEAKED: AOL's Master Plan." Business Insider, February 1, 2011. http://www.businessinsider.com/the-aol-way?op=1.

Carr, J. Frank. "The 'Quick Excerpt' Splog and Why You Shouldn't Worry." OpTempo, November 12, 2007. http://optempo.com/2007/11/12/the-quick-excerpt-splog-and-why-you-shouldnt-worry/.

Carter, Rick. "From Drawings to Sets." AI: Artificial Intelligence DVD Disc 2. Dreamworks Video, 2002.

Casey, Timothy D. *ISP Liability Survival Guide: Strategies for Managing Copyright, Spam, Cache, and Privacy Regulations*. New York: Wiley, 2000.

Cavanagh, Allison. *Sociology in the Age of the Internet*. Berkshire: Open University, 2007.

Ceruzzi, Paul E. *A History of Modern Computing*. Cambridge, Mass.: MIT Press, 2003.

Chansler, Bob. "Re: Close, but No Cigar." As quoted in Brian Reid, "MSGGROUP# 506 Message headers: a note from the grass roots." April 18, 1977. http://www.std.com/obi/Networking/archives/msggroup/.

Chapman, George. *The Works of George Chapman: Poems and Minor Translations*. London: Chatto and Windus, 1875.

Chesterton, G. K. *The Man Who Was Thursday: A Nightmare*. London: Penguin, 1986.

Chitika Insights. "The Value of Google Result Positioning" (news release). May 2010. http://insights.chitika.com/2010/the-value-of-google-result-positioning/.

Claburn, Thomas. "Internet Poker Sites Seized For Fraud." InformationWeek Government (blog), April 11, 2011. http://www.informationweek.com/news/government/policy/229401726.

Coate, John. "Cyberspace Innkeeping: Building Online Community," 1998. http://cervisa.com/innkeeping.html.

Cohen, Norm. "He Wrote 200,000 Books (but Computers Did Some of the Work)." *New York Times*, April 14, 2008.

Coleman, Gabriella. "Old and New Net Wars over Free Speech, Freedom and Secrecy; or How to Understand the Hacker and Lulz Battle against the Church of Scientology" (lecture, Institute for Public Knowledge, New York University), March 31, 2009.

CompuServe v. Cyber Promotions, Inc., 962 F. Supp. 1015 (S.D. Ohio 1997).

comScore. "comScore Releases January 2011 U.S. Search Engine Rankings" (news release). February 11, 2011. http://www.comscore.com/Press_Events/Press_Releases/2011/2/comScore_Releases_January_2011_U.S._Search_Engine_Rankings.

Cormack, Gordon, and Thomas Lynam. "Spam Corpus Creation for TREC." *Proceedings of the Second Conference on Email and Anti-Spam (CEAS)* (2005).

Cormack, Gordon, and Thomas Lynam. "TREC 2005 Spam Track Overview." *Proceedings of the Fourteenth Text REtrieval Conference (TREC)* (2005). http://trec.nist.gov/pubs/trec14/papers/SPAM.OVERVIEW.pdf.

Crispin, Mark. "MSGGROUP# 696 in Reply to Jake's Message about Advertising." May 7, 1978. http://www.std.com/obi/Networking/archives/msggroup/.

Crocker, Dave. "MSGGROUP# 004 Use of a Teleconferencing System, in Place of Net Mail." June 8, 1975. http://www.std.com/obi/Networking/archives/msggroup/.

Crocker, Stephen D. "How the Internet Got Its Rules." *New York Times*, April 6, 2009.

Crocker, Stephen D. RFC 3: "Documentation Conventions." April 1969. http://tools.ietf.org/html/rfc3.

Cukier, Wendy L., et al. "Genres of Spam: Expectations and Deceptions." *HICSS '06: Proceedings of the 39th Annual Hawaii International Conference on System Sciences* (2006).

Customer Service at Portal Communications. "A Note from Portal Regarding the 'JJ' Incident." In news.admin, misc.misc, June 1, 1988.

Customer Service at Portal Communications. "JJ's Posting." In news.admin, news.sysadmin, May 27, 1988.

Da Chronic. "AOHell v3.0 Rage Against The Machine." n.d. http://www.aolwatch.org/chronic2.htm.

Dagon, David, Cliff Zou, and Wenke Lee. "Modeling Botnet Propagation Using Time Zones." *Proceedings of the 13th Annual Network and Distributed System Security Symposium (NDSS '06)* (2006).

DataPresser. "Pricing & Signup." January 2, 2010. http://web.archive.org/web/20100102111619/https://datapresser.com/signup_options.

Davis, Joshua. "Hackers Take Down the Most Wired Country in Europe." *Wired* 15.09, August 2007. http://www.wired.com/politics/security/magazine/15-09/ff_estonia?currentPage=all.

Davis, Natalie Zemon. *The Return of Martin Guerre*. Boston: Harvard University Press, 1983.

Den Beste, Steve. "Trivia on the Net." In net.misc, March 15, 1982.

Department of Defense. "DoD Internet Host Table," v. 712, PDP-10 Software Archive, February 4, 1988. http://pdp-10.trailing-edge.com/bb-ev83b-bm/01/new-system/hosts.txt.

Deutsch, Debbie. "MSGGROUP# 684 Re: The Quasar Discussion." May 2, 1978. http://www.std.com/obi/Networking/archives/msggroup/.

Dewey, John. *The Public and Its Problems*. New York: Swallow, 1954.

Dibbell, Julian. "A Rape in Cyberspace: How an Evil Clown, a Haitian Trickster Spirit, Two Wizards, and a Cast of Dozens Turned a Database Into a Society." *The Village Voice*, December 23, 1993.

Dibbell, Julian. *My Tiny Life: Crime and Passion in a Virtual World*. New York: Henry Holt, 1998.

Dixon, Robyn. "'I Will Eat Your Dollars.'" *Los Angeles Times*, October 20, 2005.

Dyson, Freeman. "How We Know." *New York Review of Books* 58 (4) (March 10, 2011). http://www.nybooks.com/articles/archives/2011/mar/10/how-we-know/?pagination=false.

Edwards, Brent Eric. "Re: ABUSE: CHAG." In news.admin.net-abuse.email, news.admin.net-abuse.misc, July 15, 1996.

Edwards, Paul. *A Vast Machine: Computer Models, Climate Data, and the Politics of Global Warming*. Cambridge, Mass.: MIT Press, 2010.

Edwards, Paul. *The Closed World: Computers and the Politics of Discourse in Cold War America*. Cambridge, Mass.: MIT Press, 1996.

Elmer, Greg. "Robots.txt: The Politics of Search Engine Exclusion." In *The Spam Book: On Viruses, Porn, and Other Anomalies from the Dark Side of Digital Culture*, ed. Jussi Parikka and Tony D. Sampson, 217–227. Cresskill, N.J.: Hampton Press, 2008.

Enright, Brandon, et al. "Storm: When Researchers Collide." *USENIX ;login:* 33 (4) (August 2008). https://www.usenix.org/publications/login/august-2008-volume-33-number-4/storm-when-researchers-collide.

Evans, Brynn. "Ethically Improper Use of Mechanical Turk?" Flickr, August 28, 2008. http://www.flickr.com/photos/bmevans/2806469717/.

Federal Energy Regulatory Commission. "Western Energy Markets: Major Issuance on March 26, 2003—Information Released in Investigation." October 10, 2003. http://web.archive.org/web/20031011133628/http://ferc.gov/industries/electric/indus-act/wem/03-26-03-release.asp.

Feenberg, Andrew. *Alternative Modernity: The Technical Turn in Philosophy and Social Theory*. Berkeley: University of California Press, 1995.

Fetterly, Dennis, et al. "Spam, Damn Spam, and Statistics: Using Statistical Analysis to Locate Spam Web Pages." *Proceedings of the 7th International Workshop on the Web and Databases* 67 (2004): 1–6.

Finjan Software. "Finjan Discovers Compromised Business and Customer Data of 40 Top-tier Global Businesses" (news release). May 6, 2008. http://www.finjan.com/Pressrelease.aspx?id=1944&PressLan=1819&lan=3.

Fishkin, Rand. "Linkbaiting for Fun and Profit." *Search Engine Journal*, November 18, 2005. http://www.searchenginejournal.com/linkbaiting-for-fun-profit/2541/.

Fitzgerald, Dan. "AOL Gives Up Treasure Hunt." *Boston Herald*, July 24, 2007. http://bostonherald.com/business/general/view/AOL_gives_up_treasure_hunt.

Forbes, Amy Wiese. *The Satiric Decade: Satire and the Rise of Republicanism in France, 1830–1840*. Lanham, Md.: Lexington Books, 2010.

Flynn, Laurie. "'Spamming' on the Internet." *New York Times*, October 16, 1994.

Franklin, Jason, et al. "An Inquiry into the Nature and Causes of the Wealth of Internet Miscreants." *Proceedings of the ACM Conference on Computer and Communications Security (CCS)* (2007).

Fraser, Matthew. "Viral Vigilantes: The Unblinking Panopticon and the Wheelie-Bin Cat Lady." 2011. http://www.digitallymediatedsurveillance.ca/wp-content/uploads/2011/04/Fraser-Viral-vigilantes.pdf.

French, Peter A. *The Virtues of Vengeance*. 2001. Lawrence, Kans.: University Press of Kansas.

Friedman, Richard. "Re: Green Card Lottery- Final One?" In alt.internet.services, alt.tasteless, alt.cyberpunk, alt.culture.internet, news.admin.policy, April 13, 1994.

FTC. "FTC Launches Crackdown on Deceptive Junk E-mail" (news release). February 12, 2002. http://www.ftc.gov/opa/2002/02/eileenspam1.shtm.

Fuller, Matthew, and Andrew Goffey. "Toward an Evil Media Studies." In *The Spam Book: On Viruses, Porn, and Other Anomalies from the Dark Side of Digital Culture*, ed. Jussi Parikka and Tony D. Sampson, 141–159. Cresskill, N.J.: Hampton Press, 2008.

Furr, Joel. "Re: ARMM: ARMM: >>>>Ad Infinitum." In alt.fan.dick-depew, news.admin.policy, alt.folklore.computers, March 31, 1993.

Gady, Franz-Stefan. "Africa's Cyber WMD." *Foreign Policy*, March 2010. http://www.foreignpolicy.com/articles/2010/03/24/africas_cyber_wmd.

Galloway, Alexander R. "Position Paper." In *Exploring New Configurations of Network Politics*, ed. Jussi Parikka and Joss Hands. Cambridge, Mass.: ARCDigital, 2010. http://www.networkpolitics.org/request-for-comments/alexander-r-galloways-position-paper.

Galloway, Alexander R., and Eugene Thacker. "On Narcolepsy." In *The Spam Book: On Viruses, Porn, and Other Anomalies from the Dark Side of Digital Culture*, ed. Jussi Parikka and Tony D. Sampson, 251–263. Cresskill, N.J.: Hampton Press, 2008.

Galloway, Alexander R., and Eugene Thacker. *The Exploit: A Theory of Networks*. Minneapolis: University of Minnesota Press, 2007.

Gansing, Kristoffer. "The Production of Waste: Functional Trash and the Politics of Spamculture" (unpublished).

Garfield, Eugene. "Citation Indexes for Science: A New Dimension in Documentation through Association of Ideas." *Science* 122 (1955): 108–111.

Genova, Judith. "Turing's Sexual Guessing Game." *Social Epistemology* 8 (4) (1994): 313–326.

Gibson, William. *Neuromancer*. New York: Penguin, 2000.

Gibson, William. *Zero History*. New York: Putnam's, 2010.

Gillett, Walter. "bozo lawyers." In comp.lang.c++, comp.client-server, alt.sys.pc-clone.gateway2000, April 12, 1994.

Gilman, Nils, Steven Weber, and Jesse Goldhammer, eds. *Deviant Globalization: Black Market Economy in the 21st Century*. London: Continuum, 2011.

Gitelman, Lisa. *Always Already New: Media, History, and the Data of Culture*. Cambridge, Mass.: MIT Press, 2006.

Glenny, Misha. *McMafia: A Journey through the Global Criminal Underworld*. New York: Knopf, 2008.

Goodfellow, Geoffrey S. "MSGGROUP# 699 [THUERK at DEC-MARLBORO: ADRIAN@SRI-KL]." May 10, 1978. http://www.std.com/obi/Networking/archives/msggroup/.

Google Inc. Securities Exchange Commission Form 10-Q. September 30, 2009. http://investor.google.com/documents/20090930_google_10Q.html.

Graham, Paul. "A Plan for Spam." paulgraham.com, August 2002. http://www.paulgraham.com/spam.html.

Graham, Paul. "Better Bayesian Filtering." paulgraham.com, January 2003. http://www.paulgraham.com/better.html.

Graham, Paul. "Plan for Spam FAQ." paulgraham.com, n.d. http://www.paulgraham.com/spamfaq.html.

Graham, Paul. "So Far, So Good." paulgraham.com, August 2003. http://paulgraham.com/sofar.html.

Graham, Paul. "Spam Is Different." paulgraham.com, August 2002. http://www.paulgraham.com/spamdiff.html.

Graham, Paul. "Will Filters Kill Spam?" paulgraham.com, December 2002. http://www.paulgraham.com/wfks.html.

Greenberg, Andy. "Reformed Hacker Looks Back." Forbes.com, August 21, 2008. http://www.forbes.com/2008/08/21/mitnick-hackers-security-tech-security-cx_ag_0821mitnick.html.

Greene, Tim. "Storm Worm Strikes Back at Security Pros." Network World, October 24, 2007. http://www.networkworld.com/news/2007/102407-storm-worm-security.html.

Grossman, Wendy. *Net.Wars*. New York: New York University Press, 1997.

Grow, Brian, and Jason Bush. "Hacker Hunters: An Elite Force Takes on the Dark Side of Computing." *Business Week*, May 30, 2005.

Hafner, Katie. *The Well: A Story of Love, Death and Real Life in the Seminal Online Community*. New York: Carroll & Graf, 2001.

Hafner, Katie, and Matthew Lyon. *Where Wizards Stay Up Late: The Origins of the Internet*. New York: Touchstone, 1998.

Hansell, Saul. "Spam Sender Settles Case in New York." *New York Times*, July 2004: 20.

Hansen, Robert. "Conversations with a Blackhat." ha.ckers.org (blog), March 14, 2010. http://ha.ckers.org/blog/20100314/conversations-with-a-blackhat/.

Hardy, Thomas. *The Mayor of Casterbridge*. London: Ballantyne, 1887.

Harris, Oliver. *William Burroughs and the Secret of Fascination*. Carbondale: Southern Illinois University Press, 2003.

Hartcup, Guy. *Camouflage: A History of Concealment and Deception in War*. Barnsely, UK: Pen & Sword Books, 2008.

Hauben, Michael, and Ronda Hauben. "On the Early Days of Usenet: The Roots of the Cooperative Online Culture." *First Monday* 3 (8) (August 3, 1998): http://firstmonday.org/htbin/cgiwrap/bin/ojs/index.php/fm/article/view/613/534.

Hauben, Ronda. "The Evolution of Usenet News: The Poor Man's Arpanet" (lecture, Michigan Association of Computer Users in Learning), March 12, 1993. http://gos.sbc.edu/h/hauben.html.

Haugh, John F. II. "Re: HELP ME!!!" In news.admin, June 10, 1988.

Hawkesworth, James. *An Account of the Voyages Undertaken by the Order of His Present Majesty, for Making Discoveries in the Southern Hemisphere, and Successively Performed by Commodore Byron, Captain Wallis, Captain Carteret, and Captain Cook, In the Dolphin, the Swallow, and the Endeavour: Drawn up from the Journals which were kept by the several Commanders, and from the Papers of Joseph Banks, Esq.*, vol. 1. Dublin: James Williams, 1775.

Hayes, Dave. "An Alternative Primer on Net Abuse, Free Speech, and Usenet." 1996. http://www.jetcafe.org/dave/usenet/freedom.html.

Hayes, Dave. "The USENET Site of Virtue FAQ." Rev. 1.7, December 6, 1997. http://www.jetcafe.org/dave/usenet/virtue.html.

Hayles, N. Katherine. "Hyper and Deep Attention: The Generational Divide in Cognitive Modes." *Profession* (2007): 187–199.

Henderson, Austin. "MSGGROUP# 522 Re: CONTENTS OF SUBJECT FIELDS." April 25, 1977. http://www.std.com/obi/Networking/archives/msggroup/.

Hess, Elizabeth. *Yib's Guide to MOOing: Getting the Most from Virtual Communities on the Internet.* Victoria, BC, Canada: Trafford, 2003.

Hilderbrand, Lucas. *Inherent Vice: Bootleg Histories of Videotape and Copyright.* Durham, N.C.: Duke University Press, 2009.

Hofstadter, Douglas. *Gödel, Escher, Bach: An Eternal Golden Braid.* New York: Basic Books, 1999.

Hofstadter, Douglas. *Metamagical Themas: Questing for the Essence of Mind and Pattern.* New York: Basic Books, 1985.

Hofstadter, Douglas. "The Turing Test: A Coffeehouse Conversation." In *The Mind's I*, ed. Douglas Hofstadter and Daniel Dennett, 69–95. New York: Basic Books, 1981.

Hollis, Ken, et al. "alt.spam FAQ or 'Figuring out fake E-Mail & Posts.'" Rev. 20050130, January 30, 2005. http://digital.net/~gandalf/spamfaq.html.

Holt, Thomas J. "Examining Costs, Benefits, and Economics in Malware and Carding Markets" (lecture, Hackers on Planet Earth Conference, New York City), July 16, 2010.

Holt, Thomas J., and Eric Lampke. "Exploring Stolen Data Markets On-Line: Products and Market Forces." *Criminal Justice Studies* 23 (1) (2010): 33–50.

Ianelli, Nicholas, and Aaron Hackworth. "Botnets as a Vehicle for Online Crime." *International Journal of Forensic Computer Science* 2 (1) (2007): 19–39.

Jerz, Dennis G. "Somewhere Nearby Is Colossal Cave: Examining Will Crowther's Original 'Adventure' in Code and in Kentucky." *Digital Humanities Quarterly* 1 (2) (Summer 2007). http://www.digitalhumanities.org/dhq/vol/001/2/000009/000009.html.

Johnson, David R. "Due Process and Cyberjurisdiction." In *Crypto-Anarchy, Cyberstates, and Pirate Utopias*, ed. Peter Ludlow. Cambridge, Mass.: MIT Press, 2001.

Johnston, Jessica. *Technological Turf Wars: A Case Study of the Computer Antivirus Industry*. Philadelphia: Temple University Press, 2009.

Jones, Matt. "The Robot-Readable World." BERG London (blog), August 3, 2011. http://berglondon.com/blog/2011/08/03/the-robot-readable-world/.

Jones, Matt. "Sensor-Vernacular." BERG London (blog), May 13, 2011. http://berglondon.com/blog/2011/05/13/sensor-vernacular/.

Joselit, David. *Feedback: Television against Democracy*. Cambridge, Mass.: MIT Press, 2007.

Joy, Bill. Interview by Eugene Eric Kim, "The Joy of Unix." *Linux Magazine*, November 15, 1999. http://www.linux-mag.com/id/349/.

"The Junk Man." "MNS: Bury the Hatchet." BlackHatWorld, February 14, 2010. http://www.blackhatworld.com/blackhat-seo/adsense/171744-mns-bury-hatchet.html.

Kafka, Franz. *The Metamorphosis and Other Stories*. New York: Penguin, 2000.

Kanaley, Reid. "Sanford Wallace, the Spam King, Abdicates and Apologizes Online." *Philadelphia Inquirer*, April 15, 1998.

Kaplan, David E. "A Land Where Con Is King." *U.S. News & World Report* (April 2001): 29.

Kelly, Kevin. "Networked Books and Networked Reading" (lecture, Books in Browsers, The Internet Archive, San Francisco), October 27, 2011. http://www.youtube.com/watch?v=HdtXo9z7uaI.

Kelly, Kevin. "Spammer AI." The Technium (blog), May 6, 2008. http://www.kk.org/thetechnium/archives/2008/05/spammer_ai.php.

Kelty, Christopher. *Two Bits: The Cultural Significance of Free Software*. Durham, N.C.: Duke University Press, 2008.

Kendall, Lori. "Community and the Internet." In *The Handbook of Internet Studies*, ed. Robert Burnett, Mia Consalvo, and Charles Ess, 309–325. Singapore: Wiley, 2010.

Kharif, Olga. "'Likejacking': Spammers Hit Social Media." Bloomberg Businessweek, May 24, 2012. http://www.businessweek.com/articles/2012-05-24/likejacking-spammers-hit-social-media.

Kilna, Anthony. "Re: Green Card Lottery- Final One?" In alt.internet.services, alt.tasteless, alt.cyberpunk, alt.culture.internet, news.admin.policy, April 13, 1994.

King, David. *The Commissar Vanishes: The Falsification of Photographs and Art in Stalin's Russia*. New York: Metropolitan Books, 1997.

Kinsmen, Gary. "'Character Weakness' and 'Fruit Machines': Towards an Analysis of the Anti-Homosexual Security Campaign in the Canadian Civil Service." *Labour/Le Travail* 35 (Spring 1995): 133–161.

Kleinpaste, Karl. "Re: C&S Have Declared War on the Net. How to Defend the Net?" In news.admin.misc, news.admin.policy, May 11, 1994.

Kleinrock, Leonard. Interview by Charles Petrie, "Len Kleinrock on the Origins of the Internet: 'This is login.'" *Computing Now*, n.d. http://www.computer.org/portal/web/computingnow/internet40/login.

Klimt, Bryan, and Yiming Yang. "The Enron Corpus: A New Dataset for Email Classification Research." *Proceedings of the European Conference on Machine Learning* (2004).

Kostakis, Vasilis. "Identifying and Understanding the Problems of Wikipedia's Peer Governance: The Case of Inclusionists versus Deletionists." *First Monday* 15 (3) (March 1, 2010): http://firstmonday.org/htbin/cgiwrap/bin/ojs/index.php/fm/article/viewArticle/2613/2479.

Krebs, Brian. "Amazon: Hey Spammers, Get Off My Cloud!" *Washington Post*, July 1, 2008.

Krebs, Brian. "Virtual Sweatshops Defeat Bot-or-Not Tests." Krebs on Security (blog), January 9, 2012. http://krebsonsecurity.com/2012/01/virtual-sweatshops-defeat-bot-or-not-tests/.

Kreibich, Christian, et al. "On the Spam Campaign Trail." *Proceedings of the 1st Usenix Workshop on Large-Scale Exploits and Emergent Threats* (2008).

Kremen, Rachel. "'Emergent' Images to Outwit Spambots." *Technology Review* (January 2010): 6.

Kropotkin, Peter. *Anarchism: A Collection of Revolutionary Writings*. Mineola, N.Y.: Dover, 2002.

Kumar, Ravi et al. "Trawling the Web for Emerging Cyber-Communities." *Computer Networks: The International Journal of Computer and Telecommunications Networking* 31 (11–16) (1999): 1481–1493.

Kwinter, Sanford. "New Babylons: Urbanism at the End of the Millennium." *Assemblage*, no. 25 (1995): 80–81.

Lane, Frederick S., III. *Obscene Profits: The Entrepreneurs of Pornography in the Cyber Age*. New York: Routledge, 2000.

Lanham, Richard A. *The Economics of Attention: Style and Substance in the Age of Information*. Chicago: University of Chicago Press, 2006.

Larson, William L. "Re: Green Card Lottery—Final One?" In news.admin.policy, April 12, 1994.

Latour, Bruno. *Aramis, or the Love of Technology*. Cambridge, Mass.: Harvard, 1996.

Latour, Bruno. "Can We Get Our Materialism Back, Please?" *Isis* 98 (2007): 138–142.

Latour, Bruno. *Reassembling the Social: An Introduction to Actor-Network-Theory*. Oxford: Oxford University Press, 2005.

Leader, Jeffery J. "Re: I'M OUT!" In news.admin.net-abuse.email, April 11, 1998.

Leonard, Andrew. *Bots: The Origin of a New Species*. New York: Penguin, 1997.

Lessig, Lawrence. *Code and Other Laws of Cyberspace: Version 2.0*. New York: Basic, 2006.

Levchenko, Kirill, et al. "Click Trajectories: End-to-End Analysis of the Spam Value Chain." *Proceedings of the IEEE Symposium and Security and Privacy* (May 2011), 431–446.

Levy, Steven. *Hackers: Heroes of the Computer Revolution*. Sebastopol, Calif.: O'Reilly Media, 2010.

Lewis, Peter H. "Protest Halts E-Mail 'Spam'—Aurora Nissan Gives Up Its Brief Attempt to Solicit New Customers by Sending E-Mail," *Seattle Times*, May 8, 1998.

Lewis, Peter H. "Sneering at a Virtual Lynch Mob." *New York Times*, May 11, 1994.

Leyden, John. "Spam DDoS Assault Cuts Off South Pacific State." *The Register*, June 25, 2008. http://www.theregister.co.uk/2008/06/25/email_ddos/.

Le Corbusier. *Aircraft*. New York: Universe Books, 1988.

Lialina, Olia. "A Vernacular Web 2." In *Digital Folklore Reader*, ed. Olia Lialina and Dragan Espenschied, 58–69. Stuttgart: Merz Akademie, 2009.

Licklider, J. C. R., and Robert Taylor. "The Computer as a Communication Device." *Science and Technology* 76 (April 1968). Reprinted in *In Memoriam: J.C.R. Licklider 1915–1990*. Palo Alto: Digital Systems Research Corporation, 1990.

Lindqvist, Sven. *A History of Bombing*. New York: The New Press, 2003.

Lions, John. *Commentary on UNIX 6th Edition, with Source Code*. San Diego: Annabooks, 1977.

Litman, Jessica. "Stealing and Sharing." *Hastings Communication and Entertainment Law Journal* 27 (1) (2004): 1–50.

Littman, Jonathan. *The Fugitive Game: Online with Kevin Mitnick*. New York: Little, Brown and Company, 1996.

Liu, Alan. Interview by Geert Lovink. "net critique," Institute of Network Cultures, February 28, 2006. http://networkcultures.org/wpmu/geert/interview-with-alan-liu.

Liu, Dongxiao. "Human Flesh Search Engine: Is It a Next Generation Search Engine?" *3rd Communication Policy Research, South Conference*, Beijing (2008).

Lloyd, Don. "Re: Green Card Lottery—Final One?" In alt.cyberpunk, April 14, 1994.

Lovink, Geert. "The Principle of Notworking (Concepts in Critical Internet Culture)" (lecture, Hogeschool van Amsterdam), February 24, 2005.

Ludovico, Alessandro. "Spam, the Economy of Desire." Neural.it, December 1, 2005. http://www.neural.it/art/2005/12/spam_the_economy_of_desire.phtml.

Maltz, Tamir. "Customary Law and Power in Internet Communities." *Journal of Computer-Mediated Communication* 2 (1) (1996). http://jcmc.indiana.edu/vol2/issue1/custom.html.

"Man in the Wilderness." In beyond-enemy-lines archive. May 24, 2000.

Markoff, John. *What the Dormouse Said: How the Sixties Counterculture Shaped the Personal Computer Industry*. New York: Viking Penguin, 2005.

Martin, Will. "MSGGROUP# 546 ABSENTEE ADDRESSEES." May 11, 1977. http://www.std.com/obi/Networking/archives/msggroup/.

Marwick, Alice. "Status Update: Celebrity, Publicity and Self-Branding in Web 2.0" (PhD dissertation, New York University), 2010.

McCarthy, John. "MSGGROUP# 692 Reaction." May 4, 1978. http://www.std.com/obi/Networking/archives/msggroup/.

McColo Hosting Solutions. "About Company." February 9, 2008. http://web.archive.org/web/20080209180850/mccolo.com/about/.

McDermott, J. "US Postal Service Is Cracking Down on Nigerian Scams." *Wall Street Journal*, November 11, 1998.

McNeil, Joanne. "The New Aesthetic: Seeing like Digital Devices at SXSW 2012." Joannemcneil.com, March 14, 2012. http://joannemcneil.com/index.php?/talks-and-such/new-aesthetic-at-sxsw-2012/.

McWilliams, Brian. *Spam Kings: The Real Story Behind the High-Rolling Hucksters Pushing Porn, Pills, and %*@)# Enlargements*. Sebastopol, CA: O'Reilly Media, 2005.

Messaging Anti-Abuse Working Group. "Email Metrics Program: The Network Operators' Perspective. Report #15—First, Second and Third Quarter 2011." November 2011. http://www.maawg.org/sites/maawg/files/news/MAAWG_2011_Q1Q2Q3_Metrics_Report_15.pdf.

Metsis, Vangelis, et al. "Spam Filtering with Naive Bayes—Which Naive Bayes?" *Proceedings of Third Conference on E-mail and Anti-Spam* (CEAS) (2006).

"mk590." "AOL for Free?" In alt.2600, January 28, 1996.

Molloy, Judy. "Public Literature: Narratives and Narrative Structures in Lambda MOO." In *Art and Innovation: The Xerox PARC Artist-in-Residence Program*, ed. Craig Harris, 102–117. Cambridge, Mass.: MIT Press, 1999.

Moore, Sally Falk. "Epilogue: Uncertainties in Situations, Indeterminacies in Culture." In *Symbol and Politics in Communal Ideology*, ed. Sally Falk Moore and Barbara G. Myerhoff, 234–235. Ithaca: Cornell University Press, 1975.

Moran, John M. "The Day the Net Changed Forever." *Hartford Courant*, June 30, 2002.

Morris, Errol. "Did My Brother Invent E-Mail with Tom Van Vleck?" Parts 1–5, *New York Times*, June 19, 2011. http://opinionator.blogs.nytimes.com/tag/tom-van-vleck/.

Moser, Bob. "Davis Wolfgang Hawke Promotes Neo-Nazi Agenda Online, Calls Himself Future American Hitler." *Southern Poverty Law Center Intelligence Report* 111 (Fall 2003). http://www.splcenter.org/get-informed/intelligence-report/browse-all-issues/2003/fall/return-of-the-kosher-nazi.

Motoyama, Marti, et al. "Re:CAPTCHAs—Understanding CAPTCHA-Solving Services in an Economic Context." *Proceedings of the USENIX Security Symposium* (August 2010), 435–452.

Mueller, Milton L. *Ruling the Root: Internet Governance and the Taming of Cyberspace.* Cambridge, Mass.: MIT Press, 2002.

Naughton, John J. *A Brief History of the Future: From Radio Days to Internet Years in a Lifetime.* New York: Overlook, 2000.

Nazario, Jose. "Estonian DDoS Attacks—A Summary to Date." Arbor Networks, May 17, 2007. http://ddos.arbornetworks.com/2007/05/estonian-ddos-attacks-a-summary-to-date/.

Nelson, Bruce. "MSGGROUP# 569 Does It Know about Mail, Too?" May 26, 1977. http://www.std.com/obi/Networking/archives/msggroup/.

Nelson, Theodor. *Dream Machines: New Freedoms Through Computer Screens—A Minority Report.* South Bend, Ind.: Self-published, 1974.

Neuwirth, Robert. *Stealth of Nations: The Global Rise of the Informal Economy.* New York: Pantheon, 2011.

Newman, Barry. "Poor Penmanship Spells Job Security for Post Office's Scribble Specialists." *Wall Street Journal*, November 3, 2011.

Nicholson, Robert. "Re: Green Card Lottery—Final One?" In bit.listserv.vse-l, demon.local, news.admin.policy, April 12, 1994.

Nissenbaum, Helen. "Privacy as Contextual Integrity." *Washington Law Review* 79 (1) (February 2004): 119–158.

Noha, Rob "JJ." "HELP ME!!AA." In comp.cog-eng, May 23, 1988.

North, Jeanne B. and Jean Iseli. *ARPANET News* 1 (March 1973). http://archive.org/details/Issue1.

"An Old Swindle Revived." *New York Times*, March 20, 1898.

Oman, Nathan B. "Washington Supreme Court Upholds State Anti-Spamming Law." *Harvard Law Review* 115 (2002): 931–937.

Optify Research. "The Changing Face of SERPs: Organic Click Through Rate" (report). April 2011. http://www.optify.net/search-marketing/organic-click-through-rate.

Paasonen, Susanna. "Irregular Fantasies, Anomalous Uses: Pornography Spam as Boundary Work." In *The Spam Book: On Viruses, Porn, and Other Anomalies from the Dark Side of Digital Culture*, ed. Jussi Parikka and Tony D. Sampson, 165–179. Cresskill, N.J.: Hampton Press, 2008.

Palmer, Bryan D. "Discordant Music: Charivaris and Whitecapping in Nineteenth-Century North America." *Labour/Le Travail* 3 (1978): 5–62.

Pansters, Danny. "mydoom.exe decyphering?" Full Disclosure (mailing list), January 31, 2004. http://seclists.org/fulldisclosure/2004/Jan/1339.

Pappalardo, Denise, and Ellen Messmer. "Extortion via DDoS on the Rise." Network World, May 16, 2005. http://www.networkworld.com/news/2005/051605-ddos-extortion.html.

Parikka, Jussi. *Digital Contagions: A Media Archeology of Computer Viruses*. New York: Peter Lang, 2007.

Parikka, Jussi, and Tony D. Sampson. "On Anomalous Objects of Digital Culture: An Introduction." In *The Spam Book: On Viruses, Porn, and Other Anomalies from the Dark Side of Digital Culture*, ed. Jussi Parikka and Tony D. Sampson, 1–21. Cresskill, N.J.: Hampton Press, 2009.

Parikka, Jussi, and Tony D. Sampson. "Part II: Bad Objects." In *The Spam Book: On Viruses, Porn, and Other Anomalies from the Dark Side of Digital Culture*, ed. Jussi Parikka and Tony D. Sampson, 101–103. Cresskill, N.J.: Hampton Press, 2009.

Parker, Graham. *Fair Use (Notes from Spam)*. London: Book Works, 2009.

Parker, Philip M. "Method and Apparatus for Automated Authoring and Marketing." U.S. Patent 7,266,767, issued September 4, 2007.

Parry, James. "Re: 'Totally Spam? It's Lubricated.'" In rec.arts.comics.strips, alt.religion.kibology, September 2, 2003.

Perez, Sarah. "Amazon's Mechanical Turk Used for Fraudulent Activities." ReadWriteWeb, August 29, 2008. http://www.readwriteweb.com/archives/amazons_mechanical_turk_used_for_fraud.php.

Peters, Tim. "Re: [Python-checkins] python/nondist/sandbox/spambayes GBayes.py,1.7,1.8." Python-Dev, August 22, 2002. http://mail.python.org/pipermail/python-dev/2002-August/028216.html.

Pfaffenberger, Bryan. "'If I Want It, It's Okay': Usenet and the (Outer) Limits of Free Speech." *Information Society* 12 (4) (1996): 365–386.

Pfaffenberger, Bryan. "Technological Dramas." *Science, Technology and Human Values* 17 (3) (1992): 282–312.

Pfanner, Eric, and Heather Timmons. "U.K. Seeks Global Rules for Online Gambling." *New York Times*, November 2, 2006.

Phifer, Lisa. "2006 MSSP Survey, Part 6: Managed Anti-Spam and Content Filtering." ISP-Planet, December 22, 2006. http://isp-ceo.net/technology/mssp/2006/mssp6a.html.

Pickering, Andrew. *The Cybernetic Brain: Sketches of Another Future*. Chicago: University of Chicago Press, 2010.

Postel, Jon. RFC 706: "On the Junk Mail Problem." November 1975.

Postel, Jon. "MSGGROUP# 561 Comments on RFC 724." May 23, 1977. http://www.std.com/obi/Networking/archives/msggroup/.

Prasad, Sumeet. "Microsoft's CAPTCHA Revolutions Busted by Spammers—Again and Again." Websense (blog), February 15, 2009. http://securitylabs.websense.com/content/Blogs/3306.aspx.

Preston, Richard. *The Wild Trees: A Story of Passion and Daring*. New York: Random House, 2007.

Price, Monroe E., and Stefaan G. Verhulst. *Self-Regulation and the Internet*. The Hague: Kluwer Law, 2005.

Pringle, Glen, et al. "What Is a Tall Poppy among Web Pages?" *Computer Networks & ISDN Systems* 30 (1–7) (April 1, 1998): 369–377.

Reid, Brian. "MSGGROUP# 614 Fake Robot: A Call for Help." November 21, 1977. http://www.std.com/obi/Networking/archives/msggroup/.

Rennie, Jason D.M. "Improving Multi-class Text Classification with Naive Bayes" (Master's thesis, MIT), September 2001.

Rennie, Jason D. M. "ifile: An Application of Machine Learning to E-Mail Filtering." *Proceedings of the KDD-2000 Text Mining Workshop* (2000).

Rennie, Jason D. M. "ifile README 0.1a." August 3, 1996. http://people.csail.mit.edu/jrennie/ifile/old/README-0.1A.

Reuters. "WTO Confirms U.S. Loss in Internet Gambling Case." Reuters, March 30, 2007. http://uk.reuters.com/article/2007/03/30/oukin-uk-trade-gambling-wto-idUKL3047306520070330.

Rheingold, Howard. "The Tragedy of the Electronic Commons." *Tomorrow*, December 19, 1994. http://www.well.com/~hlr/tomorrow/index.html.

Rheingold, Howard. *The Virtual Community: Homesteading on the Electronic Frontier.* New York: Harper Collins, 1994.

Rheingold, Howard. "What the WELL's Rise and Fall Tell Us about Online Community." *The Atlantic*, July 6, 2012. http://www.theatlantic.com/technology/archive/12/07/what-the-wells-rise-and-fall-tell-us-about-online-community/259504/.

Ritchie, Dennis M. "The Evolution of the Unix Time-Sharing System." *AT&T Bell Laboratories Technical Journal* 63 (6, Part 2) (October 1984): 1577–1593.

Rosenbaum, H. Jon, and Peter C. Sederberg, eds. *Vigilante Politics*. Philadelphia: University of Pennsylvania Press, 1976.

Roth, Daniel. "The Answer Factory: Demand Media and the Fast, Disposable, and Profitable as Hell Media Model." *Wired* 17 (11) (November 2009). http://www.wired.com/magazine/2009/10/ff_demandmedia/.

Rowe, Brent, Dallas Wood, and Douglas Reeves. "How the Public Views Strategies Designed to Reduce the Threat of Botnets." In *Trust and Trustworthy Computing*, ed. Alessandro Acquisti, Sean W. Smith, and Ahmad-Reza Sadeghi, 337–351. Berlin: Springer, 2010.

Ryan, Johnny. *A History of the Internet and the Digital Future*. London: Reaktion, 2010.

Sahami, Mehran, et al. "A Bayesian Approach to Filtering Junk Email." *Proceedings of the AAAI Workshop on Learning for Text Categorization,* AAAI Technical Report WS-98-05 (1998).

Schwartz, John. "Preparing for a Digital Pearl Harbor." *New York Times*, June 24, 2007.

Scoblionkov, Deborah. "Senate Embraces Spam Bill." *Wired News*, June 18, 1998. http://www.wired.com/politics/law/news/1998/06/13080.

Scott, James C. *Seeing Like a State: How Certain Schemes to Improve the Human Condition Have Failed.* New Haven: Yale University Press, 1998.

Scott, Jason. "Episode 4: 'Fidonet.'" In *BBS: The Documentary*, 2005. http://archive.org/details/BBS.The.Documentary.

Scurr, Erica. "Re: Green Card Lottery- Final One?" In rec.aviation.ifr, trumpet.announce, April 13, 1994.

EC Administrative Proceeding File No. 3-10843. "Cease-and-Desist Proceedings Instituted against Rodona Garst." July 24, 2002. http://www.sec.gov/litigation/admin/33-8113.htm.

SecureList. "Email-Worm.Win32.Mydoom.a." January 27, 2004. http://www.securelist.com/en/descriptions/old22686.

Shaviro, Steven. *Doom Patrols: A Theoretical Fiction about Postmodernism.* London: Serpent's Tail, 1997.

Shoch, John, and Jon Hupp. "The 'Worm' Programs—Early Experience with a Distributed Computation." *Communications of the ACM* 25 (3) (March 1982): 172–180.

Shubik, Martin. "The Dollar Auction Game: A Paradox in Noncooperative Behavior and Escalation." *Journal of Conflict Resolution* 15 (1) (1971): 109–111.

Sibona, Christopher, and Steven Walczak. "Unfriending on Facebook: Friend Request and Online/Offline Behavior Analysis." *44th Hawaii International Conference on System Sciences, Internet and the Digital Economy Track: Social Networking and Communities* (2011), 1–10.

Simon, Herbert A. "Designing Organizations for an Information-Rich World." In *Computers, Communication, and the Public Interest*, ed. Martin Greenberger, 37–72. Baltimore: The Johns Hopkins UniversityPress, 1971.

"sine nomine." "proposed press release, 2nd draft." In news.admin.misc, June 6, 1994.

Singel, Ryan. "Curtain Call for Junk-Fax Blaster." *Wired News*, October 9, 2004. http://www.wired.com/techbiz/media/news/2004/10/65291.

Skirvin, Tim. "What Is the Format of a Cancel Message." In "Cancel Messages" FAQ, sec. II.C, ver. 1.75. http://wiki.killfile.org/projects/usenet/faqs/cancel/.

Smith, Daniel Jordan. *A Culture of Corruption: Everyday Deception and Popular Discontent in Nigeria*. Princeton: Princeton University Press, 2007.

Smith, David. "WikiLeaks Cables: Shell's Grip on Nigerian State Revealed." *The Guardian*, December 8, 2010.

Sophos Ltd. "Pitcairn Islands Relays Most Spam per Person, Reveals Sophos" (news release). March 10, 2008. http://www.sophos.com/en-us/press-office/press-releases/2008/03/pitcairn.aspx.

Sophos Threat Analyses. "W32/MyDoom-B." January 30, 2004. http://www.sophos.com/en-us/threat-center/threat-analyses/viruses-and-spyware/W32~MyDoom-B/detailed-analysis.aspx.

Spatt, Hartley S. "Postel, Jon." In *The Scribner Encyclopedia of American Lives, Vol. 5: Notable Americans Who Died between 1997 and 1999*, ed. Kenneth T. Jackson, Karen Markoe, and Arnie Markoe. New York: Scribner's, 2002.

Stallman, Richard. "MSGGROUP# 697 Some Thoughts about Advertising." May 8, 1978. http://www.std.com/obi/Networking/archives/msggroup/.

Stallman, Richard. "MSGGROUP# 698 DEC Message [VERY TASTY!]." May 9, 1978. http://www.std.com/obi/Networking/archives/msggroup/.

Stallman, Richard. "Why Schools Should Exclusively Use Free Software." GNU Project, June 10, 2012. http://www.gnu.org/education/edu-schools.html.

Stefferud, Einar, and Dave Farber. "MSGGROUP# 675 The Quasar Discussion." May 1, 1978. http://www.std.com/obi/Networking/archives/msggroup/.

Stephenson, Neal. *Anathem*. New York: HarperCollins, 2008.

Stephenson, Neal. "Cryptonomicon Spam." n.d. http://corporation9592.com/Cryptonomicon_spam.html.

Stephenson, Neal. "Mother Earth Mother Board." *Wired* 4 (12) (December 1996): 95–161.

Stiegler, Bernard. *Taking Care of Youth and the Generations*. Stanford: Stanford University Press, 2010.

Stivale, Charles J. "Spam: Heteroglossia and Harassment in Cyberspace." In *Internet Culture*, ed. David Porter, 133–144. London: Routledge, 1997.

Stross, Charles. "It's Made Out of Meat." Charlie's Diary (blog), December 12, 2010. http://www.antipope.org/charlie/blog-static/2010/12/its-made-out-of-meat.html.

Sullivan, Danny. "Death of a Meta Tag." *Search Engine Watch*, September 30, 2002. http://searchenginewatch.com/article/2066825/Death-Of-A-Meta-Tag.

Symantec.cloud. "Global Spam Drops as Rustock Botnet Is Dismantled." *March 2011 Intelligence Report*. Symantec.cloud MessageLabs Intelligence, 2011. http://www.symanteccloud.com/mlireport/MLI_2011_03_March_Final-EN.pdf.

Szpakowski, Mark. "Community Memory: 1972–1974, Berkeley and San Francisco, California." November 4, 2006. http://www.well.com/~szpak/cm/.

Taylor, Robert. Interview by William Aspray, OH154, transcript, February 28, 1989. Charles Babbage Institute, University of Minnesota, Minneapolis.

Templeton, Brad. "Problems with H.R. 1748 (the 'Smith Bill')." On behalf of the Electronic Frontier Foundation, 1997. http://www.templetons.com/brad/spam/badsmith.html.

Templeton, Brad. "Reaction to the DEC Spam of 1978." n.d. http://www.templetons.com/brad/spamreact.html.

Terranova, Tiziana. "The Bios of Attention" (lecture, Scandic Linköping Vast, Linköping, Sweden), September 7, 2010.

Thatcher, Margaret. Interview by Douglas Keay, "Aids, Education and the Year 2000!" September 23, 1987. http://www.margaretthatcher.org/speeches/displaydocument.asp?docid=106689.

Thompson, E. P. "Rough Music: Le Charivari Anglais." *Annales* 27 (1972): 285–312.

Thompson, E. P. "Rough Music Reconsidered." *Folklore* 103 (1992): 3–26.

Train, Arthur. "The Spanish Prisoner." *Cosmopolitan*, March 1910.

Tung, Liam. "Storm Worm: More Powerful than Blue Gene?" ZDNet, September 12, 2007. http://www.zdnet.com/storm-worm-more-powerful-than-blue-gene-3039289226/.

Turing, Alan. "Computing Machinery and Intelligence." *Mind* 59 (October 1950): 433–460.

Turkle, Sherry. *Life on the Screen: Identity in the Age of the Internet*. New York: Touchstone, 1997.

Turner, Fred. *From Counterculture to Cyberculture: Stewart Brand, the Whole Earth Network, and the Rise of Digital Utopianism*. Chicago: University of Chicago Press, 2006.

U.S. House of Representatives. *Netizens Protection Act of 1997*. HR 1748. 105th Cong., 1st Sess.

U.S. House of Representatives. *CAN-SPAM*. 15 U.S.C. 7701, et seq., Public Law No. 108–187, S.877.

U.S. House of Representatives. Unlawful Internet Gambling Enforcement Act of 2006. 31 USC §§ 5361–5367.

Van Vleck, Tom. "The History of Electronic Mail." multicians.org, December 20, 2010. http://www.multicians.org/thvv/mail-history.html.

Van Vleck, Tom. "The Who Command." multicians.org, August 9, 1997. http://www.multicians.org/who.html.

The Virtual Magistrate. "Virtual Magistrate Issues Its First Decision." In *Crypto-Anarchy, Cyberstates, and Pirate Utopias*, ed. Peter Ludlow, 343–345. Cambridge, Mass.: MIT Press, 2001.

Vixie, Paul, and Nick Nicholas. "How to Sue MAPS." Internet Archive, February 29, 2000. http://web.archive.org/web/20000817002153/www.mail-abuse.com/lawsuit/.

von Uexküll, Jakob. *A Foray into the Worlds of Animals and Humans, with a Theory of Meaning*. Minneapolis: University of Minnesota Press, 2010.

Waldrop, M. Mitchell. *The Dream Machine: J. C. R. Licklider and the Revolution That Made Computing Personal.* New York: Viking Penguin, 2001.

Walker, Steve. "MSGGROUP# 002 Message Group Status." June 7, 1975. http://www.std.com/obi/Networking/archives/msggroup/.

Wall, David S. *Cybercrime: The Transformation of Crime in the Information Age.* Cambridge: Polity, 2007.

Webb, Jason. "COMMUN—THE FEMINIZATION OF CYBERSPACE," appended to "Doctress Neutopia"/Libby Hubbard, "The Feminization of Cyberspace." The Electronic Frontier Foundation Archive, March 1994. http://w2.eff.org/Net_culture/Gender_issues/net_feminization.paper.

Webber, Bob. "FCC? U.S.Mail.? (Re: JJ's Revenge—Part II)." In news.admin, news.sysadmin, June 1, 1988.

Werry, Chris. "Imagined Electronic Community: Representations of Virtual Community in Contemporary Business Discourse." *First Monday* 4 (9) (September 6, 1999). http://firstmonday.org/htbin/cgiwrap/bin/ojs/index.php/fm/article/view/690/600.

White, Harrison C. *Identity and Control: A Structural Theory of Social Action.* Princeton: Princeton University Press, 1992.

Whoriskey, Peter. "Digital Deception." *Washington Post*, May 1, 2008.

Wiener, Matthew P. "Nebraska letter." In news.misc, May 27, 1988.

Wiener, Norbert. *Cybernetics: or Control and Communication in the Animal and the Machine.* Cambridge, Mass.: MIT Press, 1948.

Williams, Alex. "The Largest Cloud in the World Is Owned by a Criminal Network." ReadWriteWeb (blog), April 19, 2010. http://www.readwriteweb.com/cloud/2010/04/the-largest-cloud-in-the-world.php.

Williams, Raymond. *Keywords: A Vocabulary of Culture and Society.* New York: Oxford University Press, 1985.

Williamson, Charles W., III. "Carpet Bombing in Cyberspace: Why America Needs a Military Botnet." *Armed Forces Journal*, May 2008.

Wohl, Robert. *A Passion for Wings: Aviation and the Western Imagination 1908–1918.* New Haven: Yale University Press, 1996.

Wohl, Robert. *The Spectacle of Flight: Aviation and the Western Imagination 1920–1950*. New Haven: Yale University Press, 2007.

Wolcott, Jennifer. "You Call It Spam, They Call It a Living." *Christian Science Monitor*, March 22, 2004. http://www.csmonitor.com/2004/0322/p12s02-ussc.html.

Wu, Tim. *The Master Switch: The Rise and Fall of Information Empires*. New York: Knopf, 2010.

Yardi, Sarita, et al. "Detecting Spam in a Twitter Network." *First Monday* 15 (4) (January 2010). http://firstmonday.org/htbin/cgiwrap/bin/ojs/index.php/fm/article/view/2793/2431.

Zellich, Rich. "MSGGROUP# 693 INOVATIONS IN ENGINEERING PUBLICATION." May 4, 1978. http://www.std.com/obi/Networking/archives/msggroup/.

Zhang, Harry. "The Optimality of Naive Bayes." *Proceedings of the Seventeenth Florida Artificial Intelligence Research Society Conference* (2004).

Zielinski, Siegfried. *Deep Time of the Media: Toward an Archaeology of Hearing and Seeing by Technical Means*. Cambridge, Mass.: MIT Press, 2008.

Zipf, George K. *Human Behavior and the Principle of Least Effort: An Introduction to Human Ecology*. Cambridge, Mass.: Addison-Wesley, 1949.

Zipf, George K. *The Psycho-Biology of Language: An Introduction to Dynamic Philology*. Cambridge, Mass.: MIT, 1965.

INDEX